Liberty an
Research an
Development

The Hoover Institution
gratefully acknowledges
the support of

JOANNE AND JOHAN BLOKKER

on this project.

PHILOSOPHIC REFLECTIONS ON A FREE SOCIETY

Liberty and Research and Development

Science Funding in a Free Society

Edited by
Tibor R. Machan

HOOVER INSTITUTION PRESS

Stanford University   Stanford, California

www.hoover.org

Hoover Institution Press Publication No. 506

First printing 2002
08   07   06   05   04   03   02   9   8   7   6   5   4   3   2   1

Manufactured in the United States of America
The paper used in this publication meets the minimum requirements
of American National Standard for Information Sciences—Permanence
of Paper for Printed Library Materials, ANSI Z39.48-1984. ♾

Library of Congress Cataloging-in-Publication Data
Liberty and research and development : science funding in a free
society / edited by Tibor R. Machan
          p.   cm. — (Philosophic reflections on a free society)
(Hoover Institution Press publication ; no. 506)
    Includes bibliographical references and index.
    ISBN 0-8179-2942-8 (alk. paper)
    1. Federal aid to research—United States.   I. Machan, Tibor R.
II. Series.   III. Hoover Institution Press publication ; 506.
Q180.U5 L47   2002
338.973'06—dc21   2002024250

CONTENTS

ACKNOWLEDGMENTS

ONCE AGAIN I wish to express my gratitude to the Hoover Institution on War, Revolution and Peace, and its director, John Raisian, for supporting the publication of this work. Joanne and her late husband Johan Blokker gave generously in support of the Hoover Institution Press series, Philosophic Reflections on a Free Society, for which I wish to express my deep gratitude. The contributing authors were cooperative, patient, and conscientious throughout the entire publishing process. David M. Brown has helped with some editing and I wish to thank him for this. Tina Garcia has been very helpful with administrative aspects of this project and, once again, the diligence of Pat Baker, Ann Wood, and Marshall Blanchard of the Hoover Institution Press is also much appreciated.

CONTRIBUTORS

MICHAEL BLASGEN is a private investor and consultant in information technology. Most recently he was vice president and head, Computing Technologies Laboratory, at Sony's U.S. Research Laboratory. He was formerly director of IBM's Austin Research Laboratory that completed the world's first 1 GHz microprocessor. His prior positions include director of RISC Systems at IBM's T. J. Watson Laboratory responsible for the 801 project that led to the introduction of the RISC System/ 6000, and manager of database systems at IBM's Almaden Laboratory responsible for System R, the first relational database that led to DB2. He holds a Ph.D. in electrical engineering and computer science from the University of California, Berkeley, a B.S. from Harvey Mudd College, and an M.S.E.E. from California Institute of Technology. He is a Fellow of the Association for Computing Machinery, and a Fellow of the Institute of Electrical and Electronic Engineering.

R. PAUL DRAKE teaches and pursues research while also directing the Space Physics Research Laboratory at the University of Michigan. His current research emphasizes the laboratory simulation of astrophysical and space phenomena. His degrees include a B.A. in philosophy and physics from Vanderbilt University (1975) and a Ph.D. in physics from Johns Hopkins (1979). He has authored more than 130 scientific papers and is a Fellow of the American Physical Society.

OLIVER MAYO is a quantitative geneticist and statistician who has spent the last twelve years running Animal Production, one of the business units of the Commonwealth Scientific and Industrial Research Organisation, Australia's national research agency. His published work includes books on plant breeding, evolution, human genetics, and Australian wine. (He had a small vineyard back when he was a carefree academic at the University of Adelaide.) He is a Fellow of the Australian Academy of Science, the Australian Academy of Technological Sciences and Engineering, and a Foreign Member of the Russian Academy of Agricultural Science.

ELEFTHERIA MARATOS-FLIER is an associate professor of medicine at Harvard Medical School and head of the Section on Obesity at the Joslin Diabetes Center. Her studies involve molecular mechanisms regulating appetite and body weight (in mouse and man) and she is the author of more than forty scientific articles. She is also interested in health and science policy and medical ethics.

TIBOR R. MACHAN is Distinguished Fellow and Freedom Communications Professor of Business Ethics and Free Enterprise at the Leatherby Center for Entrepreneurship and Business Ethics, Argyros School of Business and Economics, Chapman University, and a Research Fellow at the Hoover Institution.

Some Skeptical Reflections on Research and Development

Tibor R. Machan

SHOULD GOVERNMENT be funding much of the science in a free society?

In the years since the Second World War, it has become commonplace to expect that most advanced scientific research will be funded by the federal government. Following that conflict's intensive scientific and technological demands and the federal government's efforts to meet them, most scholars concerned with research and development have approached their subject with the idea that government funding is and ought to be the norm.

Yet the U.S. government has not always funded research and development as it does now, and before the war there was little such funding. The exception was agricultural research, which—as the denizens of land-grant universities are wont to remind us—the federal government began subsidizing with the first Morrill Act in 1862.

I wish to thank Professor Robert Campbell and Michael Blasgen for their help with some of the material in this Introduction.

After the Second World War the defense establishment realized what a huge competitive advantage it had obtained by the advanced scientific work (code-breaking, computers for weapons targeting, bombs, planes) done at their behest before and during the war, and hoped to stimulate further innovation. This hope led to the Department of Defense's massive spending on research and development.

Early in the Truman administration, there was an argument that not all federal funding of research and development should be driven by the needs of the Department of Defense, and so the National Science Foundation was born. A similar argument, in the wake of the Soviet Union's launch of Sputnik, led to the establishment of the National Aeronautics and Space Administration and, later, the National Institutes of Health. Together, these organizations now plow billions into research and development each year, adding up to about one-third of all expenditure. Despite the large raw numbers, over the years the percentage of total research and development funding has actually been dropping, down from about two-thirds of the total in the 1960s. The trend will likely continue as the economy grows faster than government-funded research. Some government-sponsored research and development has doubtless been scientifically and technologically fruitful, producing a higher standard of living and a better-informed citizenry. The example most often cited is the work on data communications leading up to the Internet, originally developed by the Department of Defense and the National Science Foundation. The World Wide Web was born of research sponsored by European governments at CERN, the European Laboratory for Particle Physics. The first easily available web browser, Mosaic, was developed by the National Center for Supercomputing Applications, a spin-off of the National Science Foundation.

Despite the undoubted usefulness of some of these expenditures, there remains the question of whether the funds were necessarily deployed in *the* best way. We know how the money has been spent by the government, but we don't know how it would have been spent had it remained in private hands. Might some other projects have served more valuable purposes? And could these alternative projects have been pursued by government, or is the private sector better equipped to discern and to pursue such purposes?

When it is noted, for example, that the Internet was made possible by government support, that sheer historical fact does not settle the debate by a long shot. Consider the following: The Defense Department's role in developing ARPANET, the forerunner to the Internet, was more as a customer than as an engineer creating something by design. It provided money for researchers doing early work on a decentralized computer network, but it didn't anticipate anything like the Internet we use today. Indeed, the essentially unplanned way in which the Internet developed is an example of the biologically informed models of growth and self-regulation that libertarians celebrate. It's also worth pointing out that the Internet's huge growth, both in terms of infrastructure and customers, came about due to commercial investment, not government financing.[1]

Without a market in which allocations can be made in obedience to the law of supply and demand, it is difficult or impossible to funnel resources efficiently with respect to actual human preferences and goals.[2] One of the insights of the late

---

1. Brian Doherty, "Cybersilly," review of *Cyberselfish*, by Paulina Borsook, *Reason* (August/September 2000): 68.

2. See, however, Terrence Kealey and Aram Rudenski, "Endogenous Growth Theory for Natural Scientists," *Nature Medicine* 4, no. 6 (September 1998): 995–99, and Terrence Kealey, *The Economic Laws of Scientific*

Nobel Laureate economist F. A. Hayek, and of his teacher, Ludwig von Mises, is that central planning is not so much undesirable as impossible. The signals that communicate what needs to be done (i.e., prices) are necessarily absent: for it is precisely the free play of supply and demand in the market which generates those signals. Central planners thus lack the feedback mechanism of the price system that enables market agents to figure out what needs to be produced, how much, and of how high a quality.

Market economies have problems enough when price mechanisms are merely hampered (through price controls mandating maximum or minimum prices). When market processes are eliminated altogether, central planners do not have even mangled price information to consult. The issue isn't whether what such planners dictate is important or worthwhile, but whether, absent the data about priorities conveyed through price signals created by freely acting individuals, determinations about what is socially important can even be made at all.

Consider the immense funds used to build massive accelerators to support experiments in the field of particle physics alone. No one can dispute that there is a value—to someone— in knowing what such research produces for any of the sciences receiving it, including particle physics. No one disputes the derivative values that may emerge from such expenditures, sometimes for decades, fueling the economy and satisfying many wants and needs. But if Hayek and Mises are right, it's far from certain that those monies could not have been spent better. As central planners make decisions about how the col-

---

Research (London: Macmillan, 1996). These works argue that reasons exist to believe that the good works engendered by government funding of research and development could and would have been attained via the funding a laissez-faire approach would have produced.

lective wealth—consisting mostly of taxes—should be spent, they need a set of priorities. But how to determine them?

In a socialist conception of social life the solution is simple: a theory of society's needs guides the allocation of resources. But since the needs of society are many, highly varied, and disparate—those of the individuals of whom the society is comprised—the supposed solution is just that, merely supposed. To solve the calculation problem, central planners require much more than their theory of collective need and must also determine individual needs, wants, wishes, and preferences. This determination cannot be made without knowing what members of society would have done had they been allowed to allocate their resources in terms suited to them. (In which case, why not just let them do that?) A price system is necessary, one that helps all the members of society communicate the degree of importance resources have. As Hayek put it, "It is more than a metaphor to describe the price system as telecommunications which enables individual producers to watch merely the movements of a few pointers, as an engineer might watch the hands of a few dials, in order to adjust their activities to changes of which they may never know more than is reflected in the price movement."[3] It is the price system which makes possible the most accurate and extensive learning of what is needed in society.[4]

As with other arenas taken over by government, federally funded research and development is burdened by all the earmarks of lumbering nonprofit bureaucracies: repeated cost overruns, fuzziness about goals, inability to ensure that the purposes of the vested interests lobbying for funds are congru-

---

3. F. A. Hayek, *Individualism and Economic Order* (London: Routledge, 1949), 86–87.

4. I discuss the topic, from a normative perspective, in Tibor R. Machan, *Capitalism and Individualism* (New York: St. Martin's Press, 1990).

ent with valid public purposes, and so forth. All of these problems spawn massive misallocation of resources and lack of public confidence.

We can go further and point out that the monies are almost certainly misspent from the perspective of the individual taxpayer—who would choose to spend his money in any other way than on the specified public project if allowed to make the decision himself. Maybe he regards paying the rent or funding his children's college education to be more important than the long-run benefits of particle accelerators, or is the director of a research and development department in a private company, able to invest more funds in his firm's own research and development projects if the government did not take so much of the firm's income for its own research and development.

Consider the nature of expropriation as such. Consider all the criminals who have stolen, burglarized, and embezzled from others. They have spent what they looted on various goods and services, to fulfill purposes which they themselves undoubtedly regard as important. The expenditures of these thugs and thieves clearly contribute to the economic well-being of those who produce goods and services that they purchase as well as others down the line who receive payments from these producers. That is what we see. What we don't see is how those who rightfully owned the funds would have expended them had they not been stolen. The victims are not able to pay for their own projects. They are as deprived as are, indirectly, those from whom they would have purchased goods and services. The decisions made by the criminals, however much they may foster economic well-being for some, cannot rectify or compensate for the disruption of the life of the victim.

For those who prize the free society and are concerned about its vulnerability—with so many citizens losing sight of its importance and willing to sacrifice their liberty for illusory security, order, progress, safety, and so forth—the government's encroachment on any sphere of peaceful social life is ominous. A precedent is thereby established for even more extensive involvement in human affairs whenever an urgent need might be felt. The logical end of the road is the socialist disaster with wholesale central planning of the economy and banning of individual choices.

We know that, when free, most individuals are trustworthy enough to expend their own honestly earned funds with reasonable wisdom and prudence, to yield worthwhile results. Under freedom, the human capacity to reason is abetted by the discipline of the market, but we cannot know the specifics: how many corporations would be able to invest in research and development or even build research and development facilities of their own, nor the outcome of these efforts. In markets, the shape of things to come depends on the combined voluntary choices and serendipitous discoveries of millions of individuals. The results are unpredictable. But we can at least predict that what happens under freedom will more closely reflect what the members of society want than what happens under top-down central planning. After all, free people make their own spending decisions directly. Not all of their decisions will be for the best, but we'd be far better off under a regime of laissez-faire than under a regime of centrally planned research-by-expropriation.

Setting aside the moral issues, is there any other method of determining social needs than the free play of the market? Can ballots pinch-hit for prices? Another Nobel Laureate, Kenneth J. Arrow, has identified the paradox of democratic social choice,[5] arguing that democracy cannot yield a reliable and

---

5. Kenneth J. Arrow, *Social Choice and Individual Values*, 2d ed. (New Haven: Yale University Press, 1963).

consistent ranking of what is important in society. The contin-
uous spectacle of group-eat-group political warfare in the wel-
fare-state democracies would seem to support the observation.

What do all these considerations tell us about the success of
public policy decisions, especially at the federal level, with
respect to allocation of valued resources to research and development? What sort of research and development ought to
be undertaken? How much of it? Where? When? To get a
glimpse of the difficulties associated with this task, one could
do worse than read James D. Savage, *Funding Science in America*.[6] Savage shows how unbelievably arbitrary and confused
is the American federal government's allocation of tax funds
to university research.

Consider the consequences of outright government control
of most American universities and heavy government subsidies
to most others. Would the effects of government research
grants be the same if all universities were private, for-profit
corporations? In other words, are management incentives in a
university affected by grants in the same way that management
incentives in corporations are affected by jockeying for government protections and subsidies? It would seem that the
inability to rationally allocate government research funds
could only compound the difficulty of choosing which kinds
of teaching, research, and administration to fund that already
afflicts most universities. The proliferation of barely account-
able administrative positions is one symptom of the general
bureaucratic malaise. Free-market economists have written
remarkably little about the economics of universities, even
though many of these economists themselves work for universities, and enjoy the protection of tenure. Yet it is obviously

6. James D. Savage, *Funding Science in America* (London: Cambridge
University Press, 1999).

not in the interests of those who want to promote a free society to treat academia as an unanalyzable exception—a social force inherently and irreducibly alien to the market.

Can public funding ever be rational? Clearly, the police and military arms of governments require their own share of research and development. Forensic and military science alone require a great deal of investment in order to meet the threats of crime and foreign aggression. In a society in which almost every realm of life is subject to democratic gerrymandering, Arrow's paradox and the calculation problem of Mises and Hayek must remain intrinsic to public processes. I have argued in previous work that if we clearly conceive the nature and boundaries of the public realm as it must be constituted in a free society, allocations of public support could be achieved rationally, free of Arrow's paradox. If the public domain is adequately circumscribed, there is no need to fight the same battles repeatedly about which projects genuinely count as public ones.[7] These purposes would be stable, resting on a definite goal that has legal standing, rather than on the shifting sands of democratic maneuvering. (Such legitimate state-funded research and development has implications for peaceful technology as well. Sharing these results would be natural in a free society, although exactly how they would be shared is another issue. For example, should government sell the findings or just give them away?

The contributors to this volume explore the implications of such problems and alternatives to the current heavy reliance on government support that research and development enjoys. Not all of the writers reach the same conclusions, but they all

---

7. Tibor R. Machan, "Rational Choice and Public Affairs," chap. 1 in *Private Rights and Public Illusions* (New Brunswick, N.J.: Transaction Books; Oakland, Calif.: Independent Institute, 1995).

squarely confront the problems. And they all keep in mind that the answers we seek cannot be arrived at dogmatically, nor even, in many cases, be anticipated at all. The free society is not a closed system, and its various emerging elements may surprise us.

Just how research and development ought to be approached in a manner consistent with the principles of human liberty is a challenging question. This volume does not directly address whether research and development ought to be approached with such principles in mind except by reference to previous entries in this series that explicitly treat arguments for individual rights. It may be useful to note that although many people speculate on what the priorities of the society in which they live should be, a community that recognizes the significance of human individuality constitutionally affirms that no ranking of priorities for society can be achieved apart from the rankings possible to specific individuals.

Regardless of how enthusiastic some people may be about research and development, or the wilderness, or physical fitness or education, these are matters for people to pursue individually and of their own free will.

CHAPTER ONE

Arguments Concerning Government's Investment in Research

Oliver Mayo

ABSTRACT

The paper addresses governmental decisions to invest in the enhancement of innovation, especially in research. The example of Australia is used to show how there is a role for government that is not simply dependent on public good or market failure.

INTRODUCTION

A government can choose to invest or not in research. Since the rise of the modern nation state, all governments have chosen to invest in research; however, for much of this period, the question of whether they should has been the subject of debate, largely for political and economic reasons. I shall draw on the

I thank S. Coffey, D. Daly, M. Hirsch, J. Mayo, G. McAlpine and D. A. A. Simpson for helpful comments, and D. Byars for data. I am proud to work for the Commonwealth Scientific and Industrial Research Organisation (CSIRO), but present only my own views in this paper, not CSIRO's.

example of Australia at length, and consider only the European-American tradition to draw some conclusions about the merit of government investment and why the question is still unresolved.

In his fable *The New Atlantis* (1627), Francis Bacon looked ahead to the establishment of a government foundation: "The end of our foundation is the knowledge of causes, and secret motion of things; and the enlarging of the bounds of human empire, to the effecting of all things possible." [1] If we read Bacon's *The Advancement of Learning* also, we can see that Bacon, for all his percipience, did not envisage great corporations with the power of states, and took it for granted that the state, in the person of the monarch, would encompass the foundation.

If we look to others who have been important in developing the intellectual framework within which we now work, Adam Smith in *The Wealth of Nations* takes it for granted that the wealth generated by commerce and industry will fund research (Book I, chapter 3 and many following places).[2] While he believes that academics should be paid by their students for their lectures, he accepts without discussion a role for the state in providing the infrastructure for tertiary education. When he argued that the great English universities were too rich and failing in their duty, he did so because this made the staff idle, not because the state had no place in universities.

Thomas Jefferson, in funding the expedition of Lewis and Clark, and in many other ways, demonstrated how government might usefully fund applied research; indeed, like most of the great thinkers of the late eighteenth century he did not

1. F. Bacon, *The New Atlantis* (1627; reprinted with *The Advancement of Learning* [1605], Oxford University Press, 1906), 228.
2. A. Smith, *The Wealth of Nations* (1776; reprinted in Everyman's Library, London: J. M. Dent, 1910).

distinguish between research to understand nature and research to gain mastery over nature.

In the first half of the nineteenth century, Charles Babbage, John Henry Newman, and Wilhelm von Humboldt argued, from different points of view, that the state must support higher education and that research was an essential component of higher education. McClelland[3] provides a careful discussion of the German approach, in particular the acceptance of von Humboldt's principle that a university is a place where the teachers and the taught should live their lives in science, so that it was built into the concept of state universities in an era when only Leipzig had any capacity for competition between public and private institutions. (The German sovereign states did compete with one another to attract the best staff, Justus von Liebig's move from Hessen-Darmstadt to Bavaria being the best-known. In modern times, many governments have increased funding for research to try to reverse a real or perceived "brain drain.")

The more general argument, that innovation is necessary for economic growth of developed economies, I shall take as given. Bryant and Wells[4] is a useful source for this case.

ARGUMENTS FOR GOVERNMENT
FUNDING OF RESEARCH

Why have I laid so much emphasis on the university in the introduction? In essence because, in the European world, it was the only institution that always held some persons engaged

3. C. McClelland, *State, Society and University in Germany, 1700–1914* (Cambridge University Press, 1980).
4. K. Bryant and A. Wells, eds., *A New Economic Paradigm? Innovation-Based Evolutionary Systems*, Discussions in Science and Innovation No. 4 (Canberra: Department of Industry, Science and Resources, 1998).

in advancing our understanding of the physical world. From the time of Napoleon on, the French government established new institutions with specific applied research remits, and other governments came to do the same thing, but in essence the university was the one place where research was certain to be found, however imperfectly organized and sporadically conducted. It was for this reason that all governments in Europe and North America used universities as their prime means of advancing research in the nineteenth century.

It is noteworthy that there was a parallel development of technological research, which often produced new science, as the industrial revolution spread throughout the Western world. However, this was largely funded by the new capitalists who conducted research, though they did not call it that, because it would further their goals. In doing so, they bore out Adam Smith's expectations. They were an elite. Ideas of the universal availability of schooling for all were hardly in the air, let alone higher education for all who could benefit from it, as is the case in many countries today.

Why did governments engage in research from the eighteenth century onwards? Why did they send scientists with their best mariners on voyages of exploration? Why did they send naval expeditions to observe the transit of Venus, or to prepare better charts of southern waters? A prime motivation was to support colonization, but in a sense this was subsidiary to the idea of national security. A prize for accurate determination of longitude might further science, but it would primarily help the navy. To this day, governments fund research for defense purposes, on a prodigious scale.

Education is a more important reason for funding research, insofar as it affects the ordinary citizen in time of peace. The eighteenth- and nineteenth-century writers whom I have cited already all recognized that healthy higher education requires that at least some of the teachers be engaged in creating new knowledge, in finding out more about the world in which we live, the universe in which the world sits. I have not found a single argument against the idea that universities should conduct research, and if they are to conduct research, who will pay for it? As we evolve from primitive forms of government like monarchy, we may find that private citizens more and more expect to fund higher education. However, there is no society in the world that provides evidence that the market will fund higher education for 30 or 40 percent of any given generation. The taxpayer has to do it.

Governments also fund research to enhance wealth creation, for the general well-being of a nation. This is perhaps the most contentious kind of research funding. As Macaulay would have said, every schoolboy knows that governments cannot pick winners. Everyone can point to costly failures by government in this regard, and yet it is the mechanism used by virtually all funding bodies, whether government or private, at the level of the individual scientist or the individual project or research program: funds are provided to those who are adjudged most likely to use them well and to produce results of high quality, by the standards of the day. Nevertheless, the very idea of general taxpayer support for such research remains contentious, particularly in libertarian circles.

Market failure is the ground that is perhaps most respectable for government to occupy. If we consider agriculture, we find that its mode of organization in all of the most advanced Western societies is such that very few individual primary producers can fund their own research or have a big enough stake in its outcome to take full advantage of it. Such producers will not see it as in their own interest to contribute to the funding of research unless they can do it collectively and gain collective benefits. Accordingly, governments have historically funded a substantial quantum of agricultural research in all countries. This is changing; agribusiness now operates on such a scale that it can fund some of its own research. It is noteworthy that even where this is the case, government still funds significant research, both within government institutions and the industries themselves. It may be that, in certain economies, this particular market failure is ending. If true, it will put small producers at a bigger disadvantage than at present, because they will not be able to fund their own individually targeted research.

That something may be happening in the United States or Germany does not mean that it is happening everywhere. Consider the wool industry in Australia. This industry was established about 190 years ago, and has been successful for about 150 of those years. A century ago, it was the major factor in giving Australia the highest income per capita in the world. Eighty years ago, in a period of relatively low prosperity, the Australian government recognized that individual growers were not wealthy enough to fund the research they needed to solve general industry problems, and instituted mechanisms to solve problems on a national scale. The establishment of the Commonwealth Scientific and Industrial Research Organization (as it is today) recognized that a nation of fewer than ten million people sparsely spread over a whole continent would not find sufficient resources anywhere to address major industrial and environmental problems. What is noteworthy is that, when the government took an initiative to employ scientists to solve such problems such as parasites in sheep, wealthy individuals from the industry supported the work further financially, with the enthusiastic, informed collaboration that is the best way to take research results into agriculture. But government initiative, with a national rather than a parochial perspective, was needed in the state of development of Australia at the time.

If we look at the wool industry, we see that on-farm turnover was about AUD6000 million in 1990, and with the generally accepted means of funding—whereby graziers contributed one half of one percent of gross turnover and government matched this—there was some AUD60 million available to support research for the wool industry. However, market conditions a decade later have halved the contribution by producers and therefore the matching contribution from government, so that at a time when research is desperately needed to maintain the competitiveness of an inherently highly variable and expensive natural fiber, the funds available from the industry have in real terms dropped by over 75 percent. This is a real market failure as there are still some 40,000 wool growers and the top 20 percent do not produce 80 percent of the wool. Individual producers cannot capture the benefits of research and therefore should not be funding it for themselves; they are simply too small to fund projects that each cost one million Australian dollars or more.

If we turn to investment in research by the private sector, the reasons for such funding are rather different. First, there is benevolence, to regard it most positively. Individuals and corporate entities give money to research out of the goodness of their hearts or for general well-being. Foundations such as Rockefeller, Wellcome, and Ford are prime examples, and have unquestionably been of great benefit to humanity. However, they have not been subjected to the pressures of the market—they have, as mentioned above, worked by "picking winners." Furthermore, we need to recognize that there are moral arguments against foundations. John Stuart Mill provided some: he argued that the establishment of any perpetuity should not depend on the founder's wishes, otherwise we "make the dead, judges of the exigencies of the living." [5] Having established the impropriety of a foundation, Mill determined on utilitarian grounds that foundations are, in general, good in their effects because, without them, there will be insufficient resources for the purposes to which they are addressed. He recognized that, at any particular point in time, the market would not find sufficient resources for education, research, or any of the purposes for which foundations were established. To repeat, he first established a sound argument that perpetual foundations were morally unsound.

The prime motive of all research is curiosity: how is it that things are as they are, and how might they be if we knew how to change them? As has frequently been pointed out, Charles Darwin, who did more than anyone else except, perhaps, Copernicus and Freud to change our Western view of our place in nature, never undertook a day's paid employment in his life. Moreover, his contribution, unlike that of Freud, has stood the test of time. While few independently wealthy scientists may be pointed to today, Darwin was not the only example of his kind between 1750 and 1900. It was an era when enormous advances were produced by people whom today some sneer at as amateurs. They do so foolishly, because the love of the thing is the most important driver for progress in any field of human endeavor.

Second, the private sector funds education because it needs an educated workforce, and it pays people who will do much more research than teaching because it knows that such people are essential for healthy advanced teaching to exist; the German arguments of 200 years ago have long been settled.

Finally, and very important but not necessarily as important

5. J. S. Mill, *Early Essays*, ed. J. W. M. Gibbs (London: George Bell, 1897).

as many think, is profit. The reason for my caveat is that there are still, especially in Australia, people who regard research as a cost rather than an investment. The private sector funds an enormous quantum of research in all advanced societies, including basic research in public institutions, in order to stay in business, to grow, and to be successful. Whatever the precise role of basic research in economic growth, one has, as noted earlier, to accept that some research is needed for economic growth. At the level of the enterprise, innovation is not simply application of research, but research is a significant component of innovation, especially in new and rapidly changing industries.

Before I attempt to evaluate the merits and balance of the different arguments, I will look further at the example of Australia.

THE CASE OF AUSTRALIA

European Australia developed as a set of distinct colonies, and science and technology have always been hampered by the rivalry among these colonies. Initially, research was also hampered by the small scale of any activity compared with what could be undertaken in Europe or North America, and by the consequent lack of specialization that ensured relative inefficiency, however enterprising individual innovators might be. Despite these difficulties, research related to industry did begin before Australia became a nation, mostly to support agriculture and mining, industries in which new ideas might be imported but in which those ideas had to be tested locally. Todd[6] gives a vivid picture of the development of local technology

6. J. Todd, *Colonial Technology: Science and the Transfer of Innovation to Australia* (Cambridge University Press, 1995).

and research capacity in livestock vaccines and gold ore processing between 1880 and 1910.

The development of vaccines drew on the innovations of Louis Pasteur, which the Pasteur Institute was spreading overseas through local subsidiaries. Australians recognized the need to use these new approaches to solve the problem of anthrax in the eastern states and hoped that it might also provide an approach to the biological control of rabbits. The former was successful, and vaccines produced locally by a small private company and a small public institute displaced the semi-imported Pasteur product. As well as stimulating the development of local enterprises, this major livestock problem also led to fruitful collaboration, in at least two colonies, among state departments of agriculture, universities, graziers, and the small commercial and manufacturing sector. Indeed, the companies that were established have successors that are still in business today.

Solution of practical problems led to technological innovation and scientific advance but, perhaps crucially for the future, not to development of indigenous technologies that might generate whole new industries. Scale was a factor in this outcome, but so were the needs of derivative industries (mining, agriculture based on established livestock and crop plants) and the availability of starter technologies.

When the Commonwealth of Australia was established in 1901, it had no role in research and development, or in education. Although politicians called for establishment of a department of agriculture, with research responsibilities, as early as 1901, it was not until World War I revealed Britain's, and by extension Australia's, hazardous dependence on German technology that the establishment of a research body with a national mandate was taken seriously. This is recounted by Currie and Graham. The focus of most of the proponents was agriculture, the basis of Australia's wealth. There was widespread recognition that productivity was low, that science could increase it, and that market failure was hindering progress. But it was not all agriculture: Prime Minister William Morris Hughes announced in 1915 that a national body would be formed, "There must be a combination of science and business capacity."[7]

A body, the Commonwealth Institute of Science and Industry, was not set up until 1921, and it was not until 1926 that a truly national body was established, the Council for Scientific and Industrial Research. This occurred despite the urging of the most influential people in the community, such as Sir John Monash, one of the few generals to emerge from World War I with his reputation enhanced, a brilliant engineer who had made a substantial fortune through innovation in building materials before the war, who after the war devoted himself to public service. As president of the Australasian Association for the Advancement of Science, he said in 1924:

> The short-sighted neglect of successive Governments to make financial provision even for the bare statutory functions of the Institute have falsified the hope that under such able guidance it would become a source of varied and useful output of scientific knowledge and an inspiration to our scientific workers. We can but hope that, in course of time, a more educated public opinion will bear fruit in an adequate endowment by the State in this and in other fields of that pursuit of science which is, beyond dispute, the greatest social force in modern civilization.[8]

I am following the development of one institution because, although Australia had a proud record in innovation in agri-

7. G. Currie and J. Graham, *The Origins of CSIRO: Science and Commonwealth Government, 1901–1926* (Melbourne: CSIRO, 1966), 34.
8. Ibid., 129.

culture, mining, and mineral processing, it had a very small research effort. Universities and other tertiary institutions (schools of mines, agricultural colleges) have had a primary training role rather than a major research role. It has been estimated that in 1938 research and development constituted less than 0.1 percent of GDP (Gross Domestic Product), rising to about one percent by 1968.[9] Thus, the single institute, later council, whose expenditure is shown in Table 1, could make a significant difference to the national effort.

As the council's core funding from the treasury rose, so did the contributions from other government bodies and, more important, from industry. The council tackled major agricultural and environmental problems, which were no respecters of State boundaries. In this period, the peak level of industry funding was well over 20 percent. The Council was filling a gap; it was not crowding rural industry out of research and development; rural industry had not been able to fund its own research and development because producers were too small to capture the benefits, and overseas research and development was not always relevant to Australia's industries and environments: market failure at its simplest. From the trough of the Great Depression, the treasury appropriation rose every year, and ballooned during World War II, as the council addressed the problems of wartime production for an isolated, threatened nation. Secondary industry boomed in Australia, and technologies previously seen as too hard or too big, especially electronics and aircraft manufacture, were started. Government laboratories complemented but did not displace those developed by industry.

9. C. B. Schedvin, *Shaping Science and Industry: A History of Australia's Council for Scientific and Industrial Research, 1926–49* (Sydney: Allen and Unwin, 1987).

TABLE I. Expenditure and revenue sources, 1926/27 to 1948/49 (£ thousands)

| | | DERIVED FROM | | |
Fiscal Year	Total Expenditure	Commonwealth Treasury	Other Government Bodies	Private Industry
1926/27	45	45		
1927/28	83	82		2
1928/29	105	102		2
1929/30	142	109	30	5
1930/31	157	115	21	21
1931/32	122	70	37	15
1932/33	117	64	43	10
1933/34	125	80	29	15
1934/35	150	106	21	23
1935/36	180	122	20	37
1936/37	218	158	20	40
1937/38	264	196	19	50
1938/39	277	200	19	59
1939/40	306	232	17	57
1940/41	366	287	26	53
1941/42	431	346	18	66
1942/43	541	435	18	89
1943/44	677	561	18	98
1944/45	922	776	18	128
1945/46	1118	1009	18	91
1946/47	1505	1380	12	113
1947/48	1814	1698	14	103
1948/49	1994	1849	15	130

SOURCE: Currie and Graham 1966, Schedvin 1987.

Success of government research under the artificial condition of war prompted successive, fairly *dirigiste* governments to strengthen CSIR, renamed CSIRO, and the defense and nuclear laboratories. However, high tariff barriers and cash-starved universities did not engender innovation. Continued success for the cyclical but relatively efficient agricultural and mining sectors allowed the small economy to grow hugely for two decades after the war. The need for change did not become widely apparent for another decade, and for the last two decades of the century, successive governments tried a huge range of methods to encourage innovation.

Table 2 shows CSIRO's sources of research and development funding for the most recent years available. What happened nationally at the same time? Expenditure rose and fell in actual dollars, because of changes in business expenditure on research and development (BERD). In 1998–99, BERD was about AUD3733 million, a fall of 5 percent in the previous year, and 9 percent from the year before that. BERD represented about 0.67 percent of GDP. The peak was 0.86 percent in 1995–96.

In international comparisons, Australia is below average for Organization for Economic Cooperation and Development countries for BERD, and slightly above average for government expenditure on research and development (GERD), so that a hasty conclusion might be that government expenditure was crowding out private expenditure. However, this would be to conclude that crowding out was occurring in Australia at a total level of expenditure far below the OECD average. Table 3 shows actual expenditures and Table 4 the OECD comparative figures.

What we can see immediately from Tables 2 and 3 is that there is no detectable crowding out. Until 1997, BERD expanded faster than public expenditures on research and development (PERD), over almost a twenty-year period. During the same time, the largest single government agency expanded its external income relatively rapidly, while government funding also rose. However, once base government funds plateaued, so did external income, for which CSIRO has a target of 30 percent of total expenditure.

As noted above, BERD declined in the last year shown. This

TABLE 2. CSIRO Income from all sources 1978 to 1999
(CSIRO Annual Reports)

	Appro-pria-tion	Other Govern-ments	CRC	Other Competi-tive	Private Industry	Total
1978–79	168.9	—	—	10.6	11.6	191.0
1979–80	195.1	—	—	12.3	14.1	221.4
1980–81	248.6	—	—	15.8	18.6	283.1
1981–82	288.6	—	—	17.3	24.8	330.7
1982–83	326.6	—	—	19.0	30.7	376.3
1983–84	328.6	—	—	19.2	29.8	377.6
1984–85	319.3	—	—	20.6	40.3	380.4
1985–86	328.2	—	—	23.7	56.5	418.4
1986–87	367.9	—	—	33.2	45.0	445.9
1987–88	346.5	—	—	42.8	61.5	451.6
1988–89	348.1	—	—	54.4	69.5	472.0
1989–90	375.2	—	—	63.2	63.0	501.4
1990–91	414.3	—	—	62.2	94.5	571.1
1991–92	442.2	—	2.7	60.8	114.8	620.5
1992–93	432.6	—	8.3	82.0	175.8	698.7
1993–94	456.1	—	16.8	55.3	174.0	702.2
1994–95*	458.6	49.0	26.9	58.5	61.2	654.2
1995–96	417.8	49.8	28.4	57.5	66.4	619.9
1996–97	443.8	48.4	34.0	64.5	74.7	665.4
1997–98	417.5	52.3	32.5	63.9	77.5	693.7
1998–99	474.6	63.1	32.3	66.8	69.4	705.2

* Change in method of presentation. Competitive rural industry (wool, etc.) funds
accounted for differently.

is generally agreed to have been the result of a reduction (from
150 percent to 125 percent) of the allowable deduction of the
cost of research and development for tax purposes.[10] Whether
the decline is real may be argued for years, because advocates
of reduced government expenditure on research and develop-

10. R. Batterham, *The Chance to Change: Discussion Paper by the Chief Scientist* (Canberra: Department of Industry, Science and Resources, 2000).

TABLE 3. National Australian expenditure of research and development at current prices ($M)

	GROSS EXPENDITURE ON RESEARCH AND DEVELOPMENT		
	BERD	*Public ERD*	*GERD*
1978–79	245.8	808.8	1053.8
1981–82	373.7	1188.1	1561.8
1984–85	731.1	1684.5	2415.6
1986–87	1288.6	2085.7	3374.3
1987–88	1505.8	2229.1	3734.9
1988–89	1798.3	2482.4	4280.7
1990–91	2099.8	3122.2	5222.0
1992–93	2861.9	3621.0	6482.9
1994–95	3508.3	3958.4	7466.7
1996–97	4246.9	4557.9	8804.8
1998–99	3991.7	4858.3	8850.0

SOURCE: Supplied by Dr. Derek Byars, ABS.

ment (and tax forgone is equivalent to expenditure) consider that tax-driven research expenditure may include much creative rebadging of other expenditure. Other changes may make future increases more difficult to achieve: over the period 1996–99, the Australian Mathematical Society has shown that university teaching staff in mathematics and computing science declined by 13.5 percent, and there were smaller declines in many other sciences.

As Table 4 shows, Australia's overall expenditure on research and development is not high in OECD terms. Hence, we might not expect crowding out, if it occurs only at very high levels of government expenditure. However, since BERD is so labile in regard to incentives, it clearly can be increased through policy initiatives, just as PERD can be held static, agency by agency.

What I have not discussed, partly because of space and partly because of lack of expertise, is the relationship of national culture, character, and experience to investment that is

TABLE 4. R&D expenditure as a percentage of GDP

	AUSTRALIA			OECD
	BERD	*PERD*	*GERD*	*GERD*
1981	0.19	0.69	1.0	2.0
1985			1.1	2.3
1990	0.54	0.72	1.3	2.4
1991				2.3
1993	0.67		1.5	2.2
1995	0.73		1.6	2.2
1996			1.6	2.2
1997	0.78			2.2
1998				2.2
1999	0.71		1.6	2.2*

*Of which BERD = 1.5.
SOURCE: OECD, 2000.

inherently both risky and long-term. Australian investors in the preprivatization era always accepted risk as an essential part of investment, and in effect gambled, especially on mining exploration. There has, as yet, been no acceptance that investment in research on a large scale is a portfolio activity in which individual programs of work are no safer than a search for gold in new country, but overall the returns will be substantial. The central-limit theorem is not part of the Australian psyche.

ARGUMENTS AGAINST
GOVERNMENT FUNDING OF RESEARCH

The first argument is a moral one: government has no right to exact taxes for the purpose of funding any activity that could in principle be carried out in the private sector. Because the experiment of zero public expenditure on research has never been tried in any country that uses the results of research (one cannot count a monstrous aberration like the Khmer Republic), one can only speculate what the outcome would be. However, there is substantial evidence of market failure in many research areas, especially agriculture (as already discussed at length), preventive medicine (and public health in general) and the environment, so most people would accept that there is no absolute moral or ethical reason for government not to fund research.

Second, there is the argument that government funding of research is inherently inefficient, so that most research should not be conducted or funded by government and its agencies. This is a matter of degree; as noted, there are many examples of waste in publicly funded research and of failure of "picking winners" as a strategy, yet private nonprofit research is funded in much the same way. Accordingly, if one accepts that government has a place in funding research of some kind, concern about waste is an argument for better public accountability, rather than for zero public funding.

The set of connected arguments that is most discussed in recent years is that popularized by Terence Kealey.[11] The arguments are not wholly new, indeed perhaps date back to Joseph Priestley and Herbert Spencer, but I shall address Kealey's account because it has received a considerable degree of media credence despite some cogent criticism (see especially David,[12] Mayo,[13] and Nelson;[14] I shall only repeat material from those papers where they impinge directly on my argument).

11. T. Kealey, *The Economic Laws of Scientific Research* (Basingstoke, Eng.: Macmillan, 1996).

12. P. A. David, "From Magic Carpet to Calypso Science Policy," review of *The Economic Laws of Scientific Research*, by Terence Kealey, *Research Policy* 26: 229–55.

13. O. Mayo, review of *The Economic Laws of Scientific Research*, by Terence Kealey, *Australian Journal of Statistics* 39: 116–18.

14. R. R. Nelson, review of *The Economic Laws of Scientific Research*, by Terence Kealey, *Issues in Science and Technology* 14: 90–92.

Kealey's first argument is that the model of wealth creation through research that has been used to justify public expenditure on research is flawed. That is, he claims that policy in general is based on the idea that government-funded academic research produces outcomes that are taken up by industry in the form of technology based on the research and that this leads to economic growth. It is not clear that all governments accept this argument, but there is certainly a public "cargo cult" approach to science that matches this caricature. Kealey's view, which appears to be widely shared, is that technology is the main driver of wealth generation and demands for technology create a marketplace for research which ensures, as marketplaces are said to do, that the right research will be done. He concludes that the free market will supply sufficient basic science, by "sufficient" meaning at least as much as is supplied by any existing government policy. Private waste is irrelevant to the argument since it is private capital at risk, not the taxpayer's coerced funds.

In pursuit of this conclusion, Kealey claims to have derived a set of laws of research funding:

> The First Law of Funding for Civil Research and Development states that the percentage of national GDP spent increases with national GDP per capita.
>
> The Second Law of Funding for Civil Research and Development states that public and private funding displace each other.
>
> The Third Law of Funding for Civil Research and Development states that the public and private displacements are not equal: public funds displace more than they do themselves provide.[15]

Laws of this kind are empirical generalizations from large bodies of data. They have been very useful in science as it

15. Kealey, 245.

becomes more quantitative—for example, Dollo's Law from biology. In general terms, it states that evolution is irreversible. In the form that Dollo stated the law in 1896, it was more precise and explicit, and counterexamples were rapidly discovered. This is how science advances; a generalization is made, tested and, if not falsified, is usefully predictive. Theory is developed that has the law as a deductive outcome. We can now explain how it is that organs lost in an evolutionary lineage are never regained in that lineage, and that is how Dollo's Law is remembered today.[16]

As yet we have no theory capable of yielding Kealey's laws, because we are at the stage of testing it empirically. The evidence for the first law suggests that it is approximately true, but that the form of the relationship is not straightforward. David has suggested that the observed relationship arises partly through invalid comparison of a group of poor countries and rich countries. It may be true over time within one country but differ in functional form among countries.

The Australian evidence tabulated above forms a counterexample to the universality of application of the second law and the third law.

One particular concern for a small country that might try to develop a quantitatively rigorous science funding policy is that the laws do not address qualitative issues. From the Australian experience described above, such issues include the absence of an ordnance industry and its technological underpinnings in time of war, and the failure of native innovators to develop novel technologies that could lead to whole new industries. (Other small countries such as Holland and Switzerland have done so.) These laws could be regarded, if correct,

16. G. G. Simpson, *The Major Features of Evolution* (New York: Columbia University Press, 1953).

as addressing some of David's important questions for policy on government investment in research:

1. How closely linked are basic science and innovation in the high-technology sector?

2. Is there a balance of basic and applied research? If so, is it constant and universal?

3. How much of the GDP should be spent on research and development?

4. Within that proportion of GDP, how much public, how much private funding?

5. Should there be financial incentives for private-sector investment in research and development?

6. How effective are such incentives?

7. How well do Kealey's laws stand up to scrutiny?

Because of Australia's failure to develop novel industries, despite strong basic research, the answer to (1) concludes that there can be no inevitable commercial or industrial outcome from basic research. Similarly, (2)'s answer must involve recognition that one size cannot fit all. It is clear that the answer to (6) is that these incentives can be very effective within the levels of funding that Australia has achieved to date. If Australia's funding is perceived to be inadequate (a value judgment that I share), then the answer to (5), for Australia, is *yes*. In the same vein, for Australia the answer to (3) is that Australia should spend substantially more, perhaps up to the OECD average. (Australia's attractiveness to potential immigrants, customers and investors depends partly on the quality of its infrastructure, which is an argument against below-average performance, if nothing more.)

The answer to (7), overall, is that the laws do not stand up very well to careful scrutiny, but the ideas inherent in the laws have substantial validity. That is, governments need to justify carefully any discretionary expenditure, and watch for any disruption to effective market mechanisms, while at the same time not closing their minds to real cases of market failure and the need to initiate new policy, rather than follow trends. With these ideas in mind, let us consider how one small country, Australia, might improve its innovation.

Australia's chief scientist (Batterham) has made a number of suggestions to increase innovation and wealth-creation. The ideas, possibly good in each particular case, may be summed up as "more is better." No quantitative relationship between research expenditure and wealth is proposed; no awareness of issues such as those dramatized by Kealey informs the report. Why provide five hundred special research scholarships through the Australian Research Council (the main general federal government body supporting basic research in universities), not one thousand or some other number? Why double, not increase by say 50 percent, the number of post-doctoral fellowships? Why expand the Co-operative Research Centre Program (a scheme designed to bring research providers and industry together to conduct research and education directly relevant to industry) to involve small companies, rather than introduce competitive research and development grants available only to small companies working with research providers? Such grants would be more efficient for the funding body to administer and easier for small companies to use effectively.

Following the production of this discussion paper, a major national conference made some related recommendations:

1. Public understanding of the need for improved innovation to support national well-being and competitiveness should be facilitated through a public aware-

ness campaign, improved education for innovation, and various specific programs to foster and encourage innovation by the young.

2. Industry should receive increased tax incentives, both at a basic level (up from 125 percent to 130 percent tax deduction) and if certain targets were achieved at a higher level (200 percent).

3. Industry should be able to compete for special grants for research infrastructure.

4. Funding should be doubled for a government scheme designed to achieve rapid commercialization of new technologies.

5. Funding for the ARC should be doubled over five years.

6. Co-operative Research Centres should be encouraged to focus internationally.

7. Funding should be provided to improve international mobility of researchers to and from Australia.

These proposals were costed and would in toto draw Australia closer to OECD averages for BERD and GERD.[17] However, there was still no quantitative rationale for the differential allocation of funds.

The Australian government's response was to implement (4) and (5) as stated, and to introduce a range of measures covering all other points in varying degree (Anon., 2001). However, the quantitative differences from the recommendations and the actual funding levels chosen were not explained or justified. In toto, the new proposals represent approxi-

17. *Innovation Unlocking the Future*, Final Report of the Innovation Summit Implementation Group, Canberra, 2000.

mately one-third of the increase that would be needed to bring Australia's GERD up to the OECD average. If the increases in GERD stimulate BERD substantially, then GERD may rise more, but no predictions have been presented publicly.

If a case is to be made for a change in government funding of research, whether upwards or downwards, it needs to be based on better arguments than we find to date.

REFERENCES

Backing Australia's Ability: An Innovation Action Plan for the Future. Canberra: Dept. of Industry, Science and Resources, 2001. (ISBN 0 642 72133 5)

Bacon, F. *The Advancement of Learning*. 1605. Reprinted with *The New Atlantis*, Oxford: Oxford University Press, 1906.

Bacon, F. *The New Atlantis*. 1627. Reprinted with *The Advancement of Learning*, Oxford: Oxford University Press, 1906.

Batterham, R. *The Chance to Change: Discussion Paper by the Chief Scientist*. Canberra: Department of Industry, Science and Resources, 2000.

Bryant, K., and A. Wells, eds. *A New Economic Paradigm? Innovation-based Evolutionary Systems*, Discussions in Science and Innovation No. 4. Canberra: Dept. of Industry, Science and Resources, 1998.

Currie, G., and J. Graham. *The Origins of CSIRO: Science and Commonwealth Government 1901–1926*. Melbourne: CSIRO, 1966, 34.

Green, K. "Research funding in Australia: from the North." *Prometheus*, 4 (1986): 68–92.

Innovation Unlocking the Future. Final Report of the Innovation Summit Implementation Group. Canberra, 2000. (ISBN 0 642 72070 3)

Kealey, T. *The Economic Laws of Scientific Research*. Basingstoke, Eng.; New York: St. Martin's Press, 1996.

Koelima, J. B. J. "The Funding of Universities in the Netherlands," *Higher Education*, 35 (1998): 127–41.

Mayo, O. Review of *The Economic Laws of Scientific Research*, by Terence Kealey. *Australian Journal of Statistics* 39 (1997): 116–18.

McClelland, C. *State, Society and University in Germany, 1700–1914*. Cambridge: Cambridge University Press, 1980.

Mill, J. S. *Early Essays*. Ed. J. W. M. Gibbs. London: George Bell, 1897.

Nelson, R. R. Review of *The Economic Laws of Scientific Research*, by Terence Kealey. *Issues in Science and Technology* 14 (1997): 90–92.

Newman, J. H. *The Idea of a University*. Ed. F. M. Turner. 1873. Reprint, New Haven: Yale University Press, 1996.

OECD. *OECD Science, Technology and Industry Outlook*. Paris: OECD, 2000.

Ridler, N. B. "Aquaculture and the Role of Government Funding." *Aquaculture Europe* 22 (1998): 1–7.

Schedvin, C. B. *Shaping Science and Industry: A History of Australia's Council for Scientific and Industrial Research, 1926–49*. Sydney: Allen and Unwin, 1987.

Simpson, G. G. *The Major Features of Evolution*. New York: Columbia University Press, 1953.

Smith, A. *The Wealth of Nations*. 1776. Reprinted in Everyman's Library, London: J. M. Dent, 1910.

Spencer, H. "Specialized Administration." *Fortnightly Review* (1841). Reprinted in *Essays: Scientific, Political and Speculative*, vol. 3, 4th ed. London: Williams and Norgate, 1888, 125–70.

Todd, J. *Colonial Technology: Science and the Transfer of Innovation to Australia*. Cambridge: Cambridge University Press, 1995.

CHAPTER TWO

The Limits of Government-Funded Research: What Should They Be?

R. Paul Drake

INTRODUCTION

The United States government today funds a wide range of research, research that serves many ends. Government-funded research seeks to find the limits of the universe, to understand human health, to discover cures for all known diseases, to know the smallest constituents of matter, to improve crops, to develop new sources of energy, to maintain control of our nuclear weapons, and to develop new systems for conventional defense. This, of course, is a tiny sample of (in 1998[1]) a 67-billion-dollar enterprise. Our task here is to ask, from a classical-liberal political perspective, just how broad government support of research should be.

In recent decades libertarians and conservatives have had some success in their efforts to limit the scope of government. They have made substantial progress in deregulating some industries and in freeing some markets, but the scope of federal

---

1. National Science Board, *Science and Engineering Indicators—2000* (Arlington, Va.: National Science Foundation, 2000), chap. 2.

research funding has been almost unchanged. Many traditional conservatives are content to see the government play a substantial role in research and are often willing to support any research that does not immediately produce a profitable product. Their guiding principle, in effect, is that government must not compete with business, but can do things that support future business activity. Such conservatives would support the General Welfare clause in the U.S. Constitution.

In contrast, libertarians oppose the General Welfare clause, and do not consider it proper for the government to support future business activity. They often state their blanket opposition to government support of research. Emotional fuel for this position can be found in the work of the author Ayn Rand who, in her influential novel *Atlas Shrugged*, makes an emotionally powerful case against state control of science. She shows how the decision of a brilliant physicist to support such control leads to the destruction of his integrity, and eventually of his life, when he must progressively compromise his principles to placate those who control funding for his research. There has been a small libertarian assault on the idea that research must and should be funded by government, usually in response to arguments by prominent scientists that such funding is essential to continued economic progress and to intellectual well-being. Here again, what is at issue is the role of government in supporting the general welfare.[2]

What has been missing, however, is an analysis of government support of research that considers the purposes of government within a classical-liberal political context. In this con-

---

2. Since one does not see legal challenges to the support of research by the U.S. government, it is somewhat speculative to ground such support exclusively in the General Welfare clause. For some research, in particular as it relates to regulatory actions, the Interstate Commerce clause might also be used.

text, government exists for certain fundamental purposes, and must act to achieve these purposes. In principle, it might or might not prove necessary for the government to support research in order to achieve these fundamental purposes. We will see that it does, or can, prove necessary and this will lead us to ask what the limits of such government support should be. This question, it turns out, does not have a simple answer, but we can and will make progress toward answering it. As we explore these issues, we will review some of the relevant history and consider specific cases.

ACHIEVING FUNDAMENTAL
PURPOSES OF GOVERNMENT

We will consider research from the perspective of a classical-liberal, or libertarian,[3] view of politics, and will evaluate government actions from this perspective. The classical liberal holds that the fundamental purpose of government is to defend the natural rights of citizens. One can specify these rights, on a general level, as the rights to life, liberty, and property. These rights imply other, more specific rights, such as freedom of speech, but we will not concern ourselves here with their elucidation. The question is what the government must do in order to defend such rights. One can readily identify the need for a national defense, to protect the citizens from the potential actions of other countries, and for a police and criminal court system, to protect the citizens from criminals.

---

3. It would not be productive to engage in the largely semantic discussion regarding any differences between "classical-liberal" and "libertarian" politics. There are, for example, anarchists who consider themselves to be "libertarians" politically. The text defines the political context that will form the basis of the present discussion. Those who would develop an alternative discussion, in an alternative context, are welcome to do so.

Beyond this, the government must do other things to protect the rights of citizens and to allow citizens to exercise their rights. Here we cite a few examples but do not provide an exhaustive list. Government must protect citizens from the inadvertent violation of their rights, which might occur, for example, through the underground seeping of sewage from one property to another. It must also define the rights associated with new technologies and with intellectual property. Maintaining a patent office and patent law is one example of this. It must provide a framework for the exercise of the rights of citizens, such as for the exchange of goods and services among people. This occurs, for example, through the definition and enforcement of law regarding contracts.

One sees here that government has a number of valid purposes, but these do not imply anything about the methods through which it is to achieve them. There is a basic logical requirement that government must not achieve these purposes by routinely violating those rights of the citizens that government exists to defend. For example, the use of slave labor would not be a legitimate method by which to achieve the proper purposes of government. There are also the challenging issues of funding a government that violates no rights, and the thorny quagmire of special emergencies, which lie outside our focus here. The important point for our purposes is that there are no fundamental restrictions on the employment of individuals in order to achieve the legitimate purposes of government.

Thus, employing soldiers or policemen is valid, but so is hiring a construction company to build a jail. If a courthouse or a jail is needed, it is legitimate for the government to build it. If the government needs a steady effort in the construction of facilities, it is equally legitimate for the government to establish an internal construction organization to carry out such work. However, this is not wise, as it is very well established by political science that governmental organizations tend to be inefficient or incompetent in carrying out such activities. The wave of privatization that has been sweeping the world during the past two decades reflects an increasing recognition of this fact. The distinction drawn here will also be important for our discussion of research. I will argue that although it is legitimate for the government to pursue a wide range of research through many methods of support, it is only wise to pursue a narrower range of research and to employ specific methods of support.

There is another point worth making here. In the context of government support, there is nothing special about scientific research. Scientific research is no more and no less than one type of human activity. In some contexts, in the face of severe crises that can only be met by new knowledge, scientific research may be a very important human activity, but there are other important human activities. Those fundamental considerations that apply to government support of scientific research apply equally well to government support of any other human activity, which makes our analysis a bit easier, as we can draw examples from other areas.

From the above, one can conclude that it would be legitimate for the government to support any research required to achieve the valid purposes of government. One can ask: Is scientific research necessary for the government to achieve its proper ends? The answer is clearly yes, it can be. The easiest examples are from wartime. In the course of World War II the U.S. government supported a great deal of research that had enormous consequences for the war. Here are two of many examples. The development of radar provided advance warning of airborne attacks.[4] The development of nuclear weapons saved an enormous number of American (and Japanese) lives.[5]

Some examples from the Cold War are also compelling. Continued research in nuclear weapons and related sciences, in the face of the threat from the Soviet Union, was obviously called for by the mission of national defense. It was essential to be as capable as, if not more capable than, a nation whose explicit political philosophy involved seeking worldwide domination. For the same reasons, research that led to the development of improved technologies for spying was important. In addition, the initial steps into outer space were essential to the national defense.

The Cold War provides other cases, though, which are more difficult to evaluate. Research in the universities is a good example. From 1958 to 1968, federal support of university research increased fivefold, from $1 billion to $5 billion (in 1988 dollars).[6] On the one hand, perhaps this expansion was part of the general expansion of government involvement in life and in the economy during the 1960s. On the other hand, there were very real national-defense motivations for it. There was a need for the United States to be strong and capable in the face of the Soviet threat. Our question about this explosion of university research will be whether it was legitimate, and, if so, whether it was wise, but we are not quite ready to answer this. We first need to return to the fundamental line of argument.

4. Robert Buderi, *The Invention That Changed the World: How a Small Group of Radar Pioneers Won the Second World War and Launched a Technological Revolution* (New York: Simon and Schuster, 1996).

5. Richard Rhodes, *The Making of the Atomic Bomb* (New York: Simon and Schuster, 1986), 687.

6. Hugh Davis Graham and Nancy Diamond, *The Rise of American Research Universities* (Baltimore: Johns Hopkins University Press, 1997), 83.

CAN WE FIND LIMITS OF LEGITIMACY?

We have established that scientific research can be necessary for the government to achieve its proper ends. Such research is then legitimate. Next one would like to understand the limits of legitimacy. In particular, it would be convenient if there were a simple principle that could identify some research as legitimate and other research as not legitimate. In pursuit of this, let us examine some limiting cases. For this purpose, we can draw a useful distinction between research that directly supports actions necessary to achieve the purposes of government and research that indirectly supports such actions. Examples of research that directly supports valid governmental actions include research into the properties of high explosives and research to improve fingerprint detection. It is fairly easy to establish whether such a direct connection exists. This defines a restrictive limiting case; we will examine it first by considering this proposed simple rule: the only scientific research for which government support is legitimate is research that directly supports actions necessary to achieve the valid purposes of government. Such research would support national defense activities such as the development of weapons, or would serve a similarly narrow function with regard to other valid purposes of government. This simple rule, unfortunately, does not stand up to scrutiny. It is not hard to find cases in which much more than this would clearly be legitimate, by the standard we specified above.

One such example, which might involve some support for research but primarily involves education, is the case of a newly free third-world country. Imagine such a country, now devoted to the protection of the rights of its citizens, ready to let the magic of the market make its citizens rich. The country in question would have a substantial national defense problem.

As it became wealthier, it would become a tempting target for the poorer countries around it. Such a country would have an immediate crucial need to produce educated, technically trained individuals to support its national defense. This need would be similar to, but would go far beyond, the need for soldiers. It would clearly be legitimate for this country to provide extensive support for the general and technical education of the people it needs. This could legitimately and sensibly include establishing schools as well as sending people to Western universities. The wise pursuit of these methods might involve a delimited way for those educated to repay the government, similar to the Reserve Officer Training Corps in the United States. It might also involve a plan to phase out the government involvement in the universities as the populace became better-educated and more naturally able to provide the required support for the national defense. But the legitimate actions of government in this case, necessary for the national defense, would unquestionably include support for activities that only indirectly support national defense.

We can draw another example from the present situation of the United States. The world includes many countries with nuclear weapons, and their number will only increase with time. All such countries pose a potential threat of nuclear damage to the United States, whether deliberately or through terrorist intervention. Certain countries have the potential to become a strategic threat to the United States. The United States itself has a large number of aging nuclear weapons. For all of these reasons, the United States needs to sustain technical expertise in nuclear weapons and related areas of science and technology. However, this involves arcane subjects such as the behavior of matter at extremely high pressures and the behavior of radiation in highly ionized metals. The U.S. economy does not naturally produce experts in these areas. Moreover, within the context of science in the United States, the only way to attract sufficiently capable people into this area is to have a basic research presence in the universities. (Basic research in the universities is most often supported, not for the knowledge it produces, but for the people trained through it.) This combination of circumstances makes it legitimate and appropriate for the U.S. government to support basic research in technically relevant areas, in the universities as part of its effort to sustain technical excellence in nuclear weapons. It is worth noting that a significant fraction of astrophysics research can be justified in this way.

Thus, a narrow principle claiming that legitimate support for scientific research must directly support valid governmental action is inadequate. There are cases in which indirect needs must be met in order to achieve the valid purposes of government. However, having opened the door to indirect needs, it is unclear how far one can go. A very wide range of research activity can serve to indirectly support the valid actions of government. Examples whose connection to such valid actions is close include research aimed at developing useful technologies and research conducted to train people. Examples whose connection is real but more distant include research in improved computer architectures and basic research in any area, as history shows that one cannot predict which basic research will produce an important breakthrough for defense.

There is no reason to stop at this point in drawing indirect connections. An example whose connection is even more distant would be research into the sources of happiness (happy people are more productive, and a more productive society is more defensible). This is specific to research, but one could make similar arguments relating to a very wide range of human activity. A productive society of happy individuals can more readily defend itself. In effect, one could drive all the activity justified by the entire General Welfare clause through the tunnel created by the national defense mission. This might be an improvement as compared to the present situation, in which promoting the general welfare is explicitly identified as a purpose of government, but it is not at all where a classical liberal argument would be expected to lead.

Returning to research, in the discussion above our effort to find a simple principle that sets the limits of legitimacy has failed. There appears to be no inherent limit to the range of scientific research that the government could legitimately support beyond those that apply to all government action. Having failed to find such simple rules, we must seek wisdom. More formally, we should try to identify the contextual limits to government support of scientific research that are required by the practice of good government. This is a much harder task, and we will not come close to completing it. In particular, much of the necessary analysis in political science probably has yet to be undertaken. We can, however, make some progress.

THE PRINCIPLES OF WISDOM

Our goal here is to identify principles that provide a guide to wise government action. These will not be the sort of nearly inviolable principles that state the fundamental purposes and limitations of governmental actions but rather will be "rules of thumb," identifying the best course in the absence of compelling reasons to do otherwise. Technically, such principles can be described as *ceteris paribus* principles, which are principles that apply with "other things constant," and thus are subject to context. Though they are not absolute, these principles are tremendously important. Failing to adhere to them opens the door to many kinds of corruption and to the growth of vast, unmanageable bureaucracies that are very hard to kill. The consequences can be much greater than those involved in correcting a modest error regarding the proper limits of government action. Here is the list of principles that we will discuss and apply below.

1. Minimize government action.
2. Avoid governmental competition with private efforts.
3. Avoid sinecures.
4. Privatize as much as possible.
5. Avoid political control in awarding government funds.
6. Avoid political influence on research outcomes.
7. Specify only deliverables; avoid government involvement in process.

I make no claim that this list is complete, and some important principles may not be included. Indeed, my own view is that there is a severe need for thorough, scholarly work in classical-liberal political science. While there is substantial agreement regarding the fundamental natural rights, and that "to secure these rights, governments are instituted among men," there is a staggering absence of fundamental work regarding how government really ought to function. All we can do here is to briefly elucidate the significance of the principles listed above.

Minimize government action. This principle reflects the fact that it should be the actions of the citizens and not of government that secure the general welfare. Government has a fundamentally defensive and reactive role. It is quite clear, historically, that attempts by government to undertake organized human activities, from subways to social work, tend to produce politically driven, inefficient bureaucracies. The fact that only political action, and not profitability, can restrain such efforts allows many such organizations to continue indefinitely, consuming resources far in excess of any value they produce.

Avoid governmental competition with private efforts. This principle reflects the fact that the government exists to protect private action, not to supplant it. If the government sells some product or service to the public, from research to rocket launches to lunch, this interferes with the potential for a private individual or group to form a business that provides that same product or service. In this regard, the government can act improperly in several ways. These include (a) production by the government of a needed item that is already being produced privately, (b) entering into direct competition with an existing business, and (c) selling something at below cost, since this practically precludes the development of private alternatives.

Avoid sinecures. The principle reflects the fact that the government can, but should not, create inherently unproductive circumstances. The government can place an individual or an organization in a position that is so secure that there is no motivation for productivity or efficiency. Some organizations staffed by civil service employees are of this type, since such employees effectively cannot be laid off. In science, there are several motivations beyond tomorrow's paycheck that provide an incentive toward productivity, including the competition for supplemental funding for research projects. Nonetheless, providing indefinite funding for a research program conducted by government employees within a government laboratory is a recipe for lack of focus and low productivity, because lack of focus and inefficiency will have no adverse consequences.

Privatize as much as possible. This principle could be viewed as a consequence of the first three. It reflects the fact that any activity done in a competitive market tends to be more efficient, and vastly more innovative, than it would be if done by government. As we said above, recognition of this has led to the wave of privatization that has been sweeping the world during the past two decades.

Avoid political control in awarding government funds. This principle reflects the fact that one must work to prevent political incentives from taking precedence over the accomplishment of valid ends of the government. While political action is inherent in deciding which type of research to support and how much, just as it is inherent in deciding when to build a new courthouse, it is important in all such cases to award government funds using objective standards whose purpose is to obtain quality results. The classic examples of unwise action involve Congressmen who seek big federal construction projects for their districts. In recent years, however, scientific and academic pork have become common.[7] Each year, Congress earmarks an increasing fraction of the federal scientific budget for specific institutions. This practice, discussed further below, is most definitely not wise. The opponents of earmarking typically call for the use of peer review, which has been the traditional method. Peer review is often the best approach but has limitations, although we will not examine this issue here. A more general statement of wise policy is the following: in awarding government funds, use competitive methods with review mechanisms that employ objective evaluation processes.

Avoid political influence on research outcomes. This principle reflects the fact that the only worthwhile research outcomes are objective, but that governmental officials very often have political reasons to desire specific results. Vice President

---

7. James D. Savage, *Funding Science in America* (Cambridge: Cambridge University Press, 1999).

Gore very much desired anthropogenic global warming to be a certain, looming catastrophe while President Bush would prefer that it be small or absent. This principle complements the previous one. It is necessary, but not sufficient, to keep politics out of funding decisions. Beyond this, it is essential to the integrity of the scientific process that researchers do not feel that their conclusions will impact their future viability. Regulatory agencies routinely violate this principle—as we will discuss at length later.

Specify only deliverables; avoid government involvement in process. This principle reflects the fact that government should treat those it supports as independent contractors, not as government agents. The public has a right to know a great deal about its government agents, so there is merit in close monitoring of government agents by both the government itself and the press. The result is the emergence of a bureaucracy to monitor and regulate the actions of government agents. However, suppliers of goods or services, including research, are properly treated as independent trading partners in an open market. If they are treated instead as government agents, then the result will be (and has been) large bureaucracies that stifle their efforts at substantial cost to the taxpayers.

This completes our identification and survey of the principles of wise government that are relevant to government support of research. Each of them, and doubtless others, could be the subject of a chapter-length or book-length exposition. I hope that someone will undertake this and look forward to seeing the results.

At this point we have set the stage for a more specific discussion and will begin with a discussion of basic research and university research (these are closely coupled). This will return us to discussion of the Cold War expansion in the 1960s, in the following section. Uunfortunately, we will have to ignore funding for health research, although this is an interesting issue. The degree to which government action is appropriate, in response to disease, deserves a serious analysis. On the one hand, the bubonic plague killed a large fraction of the population of Europe[8] and the potential for a similar disaster certainly exists today. This would seem to justify some level of government activity. On the other hand, the common cold seems pretty clearly beyond the classical liberal province of government. Beyond this, there are complications in this topic related to government support and government regulation of the health industry. Overall, health research and related issues is too large a can of worms for us to open here. We will leave this to others.

HISTORICAL BASIC AND UNIVERSITY RESEARCH

In the area of basic and university research, two developments prior to World War II are worth examining. The first is government funding of research during this period.[9] Before the war, the primary justifications for government-funded research in the United States were defense and commerce, although health was also significant. From our classical-liberal perspective, we approve of defense as a motivation but not of

8. Norman F. Cantor, *In the Wake of the Plague: The Black Death and the World it Made* (New York: Free Press, 2001).

9. Daniel J. Kevles, *The Physicists* (Cambridge, Mass.: Harvard University Press, 1987); James D. Savage, *Funding Science in America* (Cambridge: Cambridge University Press, 1999); A. Hunter Dupree, *Science in the Federal Government* (Baltimore: Johns Hopkins University Press, 1986); Bruce L. R. Smith, *American Science Policy Since World War II* (Washington D.C.: Brookings Institution, 1990); Harold Orlans, ed., *Science Policy and the University* (Washington, D.C.: Brookings Institution, 1968).

commerce, viewing this as outside the proper role of government. However, the methods of accomplishing funded research during this period were distinctly unwise. There was little and reluctant funding of research by nongovernment employees. By World War II, organizations such as the Weather Service, the Geological Survey, and the National Bureau of Standards had long histories. It would have been wiser to privatize these efforts in ways that were consistent with our other principles.

Beyond the direct federal effort, much of the other funding that did exist was institutional in nature. The Hatch Act of 1887 and the Smith-Lever Act of 1914 established agricultural research and agricultural extension services that involved long-term support of politically distributed institutions.[10] If you have reminisced with farmers who were active in the mid-twentieth century, you probably know that the extension services did a lot of good work. However, from a classical-liberal perspective these efforts were wrong in two respects. First, they were undertaken to support the general welfare, rather than for a purpose connected with the rights of the citizens. Second, the way they were supported was unwise. It prevented the development of private alternatives and the innovations that would have resulted.

Meanwhile, certain U.S. universities were inadvertently preparing themselves to play a much larger role in research for the nation. "On the eve of World War II, the unplanned evolution of higher education in the United States had produced a loose, sprawling, largely unregulated system that was decentralized, pluralistic, competitive, and vast."[11] This was a direct

10. James D. Savage, *Funding Science in America* (Cambridge: Cambridge University Press, 1999), 33.
11. Hugh Davis Graham and Nancy Diamond, *The Rise of American Research Universities* (Baltimore: Johns Hopkins University Press, 1997), 24.

result of decentralized market competition. Private institutions had always played a major role, and the institutions that were publicly funded by the states competed with the private institutions and one another. This remains true today. This market in higher education differed sharply from the centralized and nationally controlled system in place in Europe, where the fraction of students advancing to the universities was three times smaller at the outbreak of World War II.

Two consequences were important. First, the best U.S. institutions (about three dozen out of more than 600) resulting from this brisk competition were extremely capable. These institutions became the research-oriented graduate schools, and there was a need for them. "The sheer size of American higher education created huge communities of scientists in every discipline and field. Although much of the instruction given at American colleges and universities took place at less than an advanced level, teaching itself created secure professional employment for thousands of scientists and scholars; and training these cadres gave work of a higher order to the graduate schools."[12]

With the advent World War II, the government needed far more research than its captive scientists could provide. The federal government was forced to rely upon project grants to support the research it needed. Christian Arnold[13] reports: "At the beginning of our involvement in the war, our military technology and machinery were almost hopelessly inadequate and obsolete. Therefore the military agencies did what they had to do—they bought the information, engineering, and 'hardware' they needed by negotiating contracts with organi-

12. Roger L. Geiger, "Organized Research Units—Their Role in the Development of University Research," *Journal of Higher Education* 61 (1990): 2.

13. Christian K. Arnold, in *Science Policy and the University*, ed. Harold Orlans (Washington D.C.: Brookings Institution, 1968), 89–90, 91.

zations and individuals that seemed likely to do the best job for them." Our response is that this was wise, and was the approach that should have been used previously. Although the top universities had the intellectual resources to undertake the needed research, in the university arena this approach led to political problems after the war. The most competent institutions won the lion's share of the work, and these institutions were concentrated in only a few states.

After World War II, the relation of government to research did not return to the prewar condition. Graham and Diamond summarize the change in perspective as follows.[14] "World War II convinced American society . . . that . . . the link between research universities and the nation's economic strength and national security was too vital for the national government to leave unattended." From a classical-liberal perspective, one would challenge this assertion with regard to the nation's economic strength, which is typically hurt by federal efforts to improve it, but one would agree that national security is the responsibility of the national government. World War II provided very clear evidence that research plays a key role in national security and in the ability to win wars. Those who analyze such matters believe that "any power that lags significantly in military technology, no matter how large its military budget or how efficiently it allocates resources, is likely to be at the mercy of a more progressive enemy."[15] The experience of the Gulf War provides recent evidence in support of this statement from 1960.

There ensued a five-year battle over the method by which

14. Hugh Davis Graham and Nancy Diamond, *The Rise of American Research Universities* (Baltimore: Johns Hopkins University Press, 1997), 25.

15. Charles J. Hitch and Roland N. McKeon, *The Economics of Defense in the Nuclear Age* (Santa Monica, Calif.: RAND, 1960), 243.

the federal government should support research.[16] One group, led by Senator Kilgore from West Virginia, favored a geographically distributed approach modeled on the agricultural extension service. They were opposed by another group, led by Vannevar Bush, who was the director of the Office of Scientific Research and Development both during and after the war. Through his wartime experience, Bush[17] "grew leery of what might happen to the independence of university research if, after the war, government directly managed the conduct of academic science." He advocated the competitive distribution of research funds by individual contracts and grants. After five years, Bush and his colleagues finally won this battle, and President Truman signed legislation creating the National Science Foundation.

Equally important, though, was what happened in the interim. Several other federal agencies had filled the policy void by beginning to support research that they needed, seeking to meet their needs from the (academic and other) research marketplace, and not by means of geographic or institutional formulae. This continued through the 1960s, so that "what emerged from the 1950s and 1960s is a relatively decentralized federal science establishment, although one dominated by national security interests, in which academic research support comes in the form of project grants awarded to individual university researchers."[18] We can see here that during this period, and from the point of view of the wise conduct of government, the good guys won. Using private resources and

16. James D. Savage, *Funding Science in America* (Cambridge: Cambridge University Press, 1999), 334–35; Hugh Davis Graham and Nancy Diamond, *The Rise of American Research Universities* (Baltimore: Johns Hopkins University Press, 1997), 28–30.
17. Ibid., 35.
18. Ibid., 36.

methods that minimized political influence was a wise way to support the research that the nation chose to undertake.

THE COLD WAR AND BEYOND

This returns us to the beginning of the Cold War and to the massive expansion of federal support for scientific research during the 1960s. This expansion has been characterized as part of the general expansion of proactive government involvement in society during this period. I believe, however, that this viewpoint is incorrect. This immense expansion, with the associated increase in facilities and in number of students, was driven in large measure by the perceived need for the United States to be strong and capable in the face of the Soviet threat. This need went far beyond the need to develop specific weapons or related technologies; it reflected the need to be ready to respond to any discovery that mattered for defense, in any area of human endeavor. Research progress in psychology was potentially as important as research in explosive chemistry. In a real sense the United States was at war, and the battleground was war-related technical capability. The Soviets understood this and spoke of it. Soviet Chairman Leonid Brezhnev pointed out in the late 1970s that the "center of gravity in the competition between the two systems [U.S. and U.S.S.R.] is now to be found precisely in [the field of science and technology]."[19] Of course, such research also can be (and was) justified for the sake of the general welfare. But this was a case where the general welfare and the needs of national defense walked hand

19. Quoted in Amos A. Jordan, William J. Taylor, and Michael J. Mazarr, *American National Security* (Baltimore: Johns Hopkins University Press, 1998), 321.

in hand. It is not surprising, in this light, that the U.S. Congress found it easy to support these efforts.

Politicians will give every reason that has a constituency as grounds for their actions, but one can sometimes find the compelling reasons by seeing which events lead to changes in their actions. It is very informative to trace the parallel evolution of national security concerns[20] and of university research.[21] During the 1950s and 1960s the Soviet threat was viewed as severe and immediate. In weapons (such as hydrogen bombs) and technology (Sputnik) they appeared to be even with or ahead of the United States. It is no coincidence that the 1960s are described as the Golden Decade for university research. By the late 1960s to early 1970s, there was a consensus among U.S. security experts that the Soviet threat had been exaggerated. The congressional response with regard to university research in the 1970s caused it to be called the Stagnant Decade. The United States de-emphasized defense in many ways during this decade, while the Soviet Union used this period to accomplish a massive buildup of its military establishment. "By the early 1980s, the enormous momentum of the Soviet buildup yielded some clear elements of Soviet superiority."[22] Again the U.S. responded, with the "Reagan buildup" of the 1980s, including the Strategic Defense Initiative. It should not be a surprise that the 1980s are characterized as a second golden age for university research.

Our focus in the above discussion has been on the United

20. Ibid., chap. 16.
21. Hugh Davis Graham and Nancy Diamond, *The Rise of American Research Universities* (Baltimore: Johns Hopkins University Press, 1997).
22. Amos A. Jordan, William J. Taylor, and Michael J. Mazarr, *American National Security* (Baltimore: Johns Hopkins University Press, 1998), 345.

States. At least one analyst of worldwide research funding (Terence Kealey) believes[23] that "governments fund education, science, universities, and technology for military, not humanitarian, reasons." He and I might differ, however, over the legitimacy of this. My view is that in many cases research funding by the U.S. government, and likely by other Western governments, has been justified and reasonable for defense reasons. Even so, much of this support has been carried out unwisely, and the needs of defense are smaller now than they were through the end of the Cold War.[24]

In the United States, things changed in the early 1990s. After the collapse of the Berlin wall and the fall of the Soviet Union, broad congressional support for university research collapsed. There was a period of substantial confusion and large fluctuations as Congress wrestled with the question of whether the "peace dividend" should include a large reduction in federal support for such research. Since that time, congressional support for science has been much less certain and far more contentious. As an active scientist, I have seen an enormous increase in the level of lobbying activity by scientific organizations and in appeals to me to contact Congress in support of some objective. Science, once coddled as essential to the national defense, has become just another interest group.

23. Terence Kealy, *The Economic Laws of Scientific Research* (New York: St. Martin's Press, 1996), 118.
24. Kealy also argues, with statistical supporting evidence, that government funding of research disproportionately displaces private research funding, leading to a net decrease in research progress. Even if he were correct in this, though, there are times when the real needs of defense would be worth the cost. However, I am skeptical of his conclusion for two reasons. First, like many scientists I am skeptical of studies that turn out to support a conclusion which the author knows in advance. Second, Kealey's analysis does not allow for whether the research is funded and managed wisely. In my view this will make a difference.

Meanwhile, Congress, perceiving various kinds of value in scientific efforts, is evolving toward a system of support in which the general welfare is a much larger motivation. Simultaneously science is increasingly part of federal "pork." Earmarking of academic and other research funds so that they go to specific institutions in specific congressional districts is steadily on the increase. Such funds have increased from $17 million in 1980 to $328 million in 1996.[25] This is not surprising. As it becomes less urgent to make research progress for the sake of defense, science funding becomes more like any other goody that Congress hands out. The classical liberal assessment of this is rather negative. Science funding is becoming less legitimate as its connection with the national defense weakens, and less wise, as earmarking becomes more significant. In this context there are many examples of programs with a plausible defense justification during the Cold War for which such a justification is much harder to support now. The Internet is an excellent example. In the early 1960s, long before Al Gore "invented" the Internet, the agency then known as ARPA (the Advanced Research Projects Agency, which later added Defense to become DARPA) focused on the problem of improving the military use of computer technology.[26] Defense laboratories and the military had a serious, unmet need for what would now be known as broadband communication. At that time, there were no commercial enterprises devoted to broadband communication. The leader of this ARPA program, Dr. J.C.R. Licklider, decided that this need could most effectively be met by work within the universities. He supported the foundation of the ARPANET, which initially connected a

25. James D. Savage, *Funding Science in America* (Cambridge: Cambridge University Press, 1999), 3.
26. This history is described at *www.netvalley.com/intval.html.*

number of universities. The great breakthrough came when researchers using a computer at UCLA tried to log in to a computer at Stanford. The system crashed after two letters were transferred, but nonetheless "a revolution had begun."[27] We can be pleased that this work laid the foundations for today's Internet, which is of great benefit and commercial importance despite the recent (as of 2001) dot.com madness. Nonetheless, at the time, it made sense for the federal government to support this work.

Today there is a follow-on project, the Internet-II, through which a number of universities are receiving support to create a next-generation Internet. One probably could justify such a project for defense reasons because everybody needs more bandwidth, including the military. But today there is a vast, commercially driven, telecommunications industry. This industry is also starving for bandwidth. I do not see why, from a classical-liberal perspective, there remains a strong justification for further federal funding of Internet development.

The space program provides another example, but with a different, more nuanced outcome. In my view, there is no question that the early space program was fully justified by defense needs. The conquest of space and development of the ability to put weapons and humans in space was essential in response to Soviet activities. Even now, there are still defense-related reasons for some space activity by the government. For example, flying and improving spy satellites remains important. In addition, as mentioned above, sustaining some astrophysics research contributes to the manpower-training and basic science needs of nuclear defense activities. However, the United States and NASA have violated our principles of wisdom in several respects. As one example, NASA should have

27. *Sacramento Bee*, May 1, 1996, D1.

encouraged the development of a private launch industry, in appropriate ways, instead of developing the space shuttle.

The reader should not conclude, however, that Congress adopted a strict, classical-liberal approach to research funding throughout the Cold War, and that this has now changed. Congress has always supported the General Welfare clause (and most definitely its local welfare corollary). Support for research in agriculture, for example, has endured since the nineteenth century.

Before leaving this subject, we should discuss the funding of research in settings other than universities, such as private laboratories. From the standpoint of validity and of wisdom, there is no general reason to prefer one over the other. In practice, much of research funding (and nearly all funding of basic research) has been supported primarily for the purpose of training students. Basic research only produces results of immediate and substantial value unpredictably and sporadically, but it produces trained scientists with high reliability when it is done in a university. This is a legitimate motivation for using universities for basic research. However, whenever this motivation is not important, it would be perfectly reasonable to support such research within other organizations.

LABORATORIES FOR
LARGE-SCALE RESEARCH

We have discussed research funded in universities, and similar research funded in private laboratory environments. In addition there is a category of research projects for which universities are ill-suited. These are projects that are of a substantial size (involving more than about ten people) or are strongly interdisciplinary (involving in an essential way more than three or four scientific disciplines). They are our subject here, because a need for them can arise in many areas of research, and because the United States has handled them quite poorly. Such projects are very difficult to accomplish in a university environment and are not very appropriate for universities, as they require tight management and a programmatic focus inconsistent with the educational context of a university. It might make sense for a university to set up a laboratory that can accomplish such projects, especially if the projects tie in to the scientific and educational purposes of the university—in fact, I now run such a laboratory. But such situations are exceptional, and the need for large-scale, interdisciplinary research projects is not.

The need for such projects led to the emergence of numerous large research laboratories during the 1950s and 1960s. The motivation for their research was the same as that for research in universities, so our above conclusions apply to the legitimacy of their research. Unfortunately, the advent and growth of these laboratories was not handled wisely. Most of them are of two types, both of which are very strongly controlled by the federal government. Some, notably including most of the NASA centers and the military laboratories (some of which are much older) are federal laboratories. In these labs, all of the property is government property and all of the employees are Civil Service employees. Other laboratories, including NASA's Jet Propulsion Laboratory (JPL) and all of the Department of Energy (DOE) labs such as Lawrence Livermore, are Federally Funded Research and Development Centers (FFRDCs). In FFRDCs, nearly all of the funding is federal, but some other organization hires the employees and manages the laboratory under a contract with the federal government. The University of California, for example, manages Livermore while Cal Tech manages JPL. The managing organization is not always a university; for example, the Sandia Laboratories are managed by Lockheed Martin. A few private research centers are large enough to qualify as laboratories in this sense, including SouthWest Research Institute (which builds space hardware and conducts space missions), but these are the exception.

Both the federal laboratories and the FFRDCs embody the worst features of a bureaucratic environment. Specifically, as there are no constraints related to profit, the only fundamental motivation of any individual involved in monitoring these laboratories, or in performing support functions within them, is to avoid adverse publicity. The most secure way to accomplish this is to *do nothing* and *prevent the scientists from doing anything*. Bureaucrats and rules proliferate without limit in such an environment. As I know from personal experience and from many discussions with employees of these institutions, the scientists working there end up facing a terrible choice. On the one hand, they can try to follow the rules. They won't succeed, but they absolutely will not accomplish anything. On the other hand, they can try to do meaningful work for the country. This requires that they routinely break the bureaucratic rules. In such an environment, it becomes very difficult to retain good judgment regarding which rules matter. In my own view, both the Wen Ho Lee incident and the missing tapes incident at Los Alamos in the late 1990s were consequences of this fundamental problem.

One specific example of bureaucratic irrationality was provided when a colleague of mine was responsible for shipping some hazardous waste from the Livermore Lab to a safe disposal site. He had three choices. He could follow DOE rules in doing so, which would require violating both federal and state laws. Or he could follow federal law, which would require violating state law and DOE rules. Or he could follow state law, which would require violating federal law and DOE rules. His situation was in addition a "Catch-22." Doing nothing would have violated the laws and rules as well.

My interactions with a few Russians illustrated more generally just how irrational the environment within such laboratories has become. During the early 1990s there was a substantial migration of Russian scientists to the United States. Some of them became affiliated with the Livermore Lab, as was I at the time. These individuals, experienced with the Soviet state, often had a great deal of difficulty understanding American culture and society. In contrast, they found it very easy to understand the workings of the DOE labs. It is truly ironic that we developed a Soviet-style environment in our attempt to counter the Soviet nuclear threat.

A more chilling example of the adverse consequences of direct federal control was provided during the Clinton administration. Vice President Gore would ask leading climate scientists, such as James Hansen and Tom Wigley, to share the stage with him while he made ludicrous statements about global warming. What the press did not report, however, was that these scientists held leadership positions in federally controlled laboratories. They had no choice but to attend this event. When the boss (or the vice boss) calls, one must go. This is the one thing I have seen, in more than twenty years of involvement with federally funded science, that reminded me of the State Science Institute as it was portrayed in *Atlas Shrugged*.

It should be clear that the development of federally controlled research laboratories was extremely unwise. It violates numbers 1 through 4 and 7 of our principles of wisdom, and sometimes also 5 and 6. Any large-scale research that is legitimately required by the nation should be undertaken by private laboratories, which should work under competitively awarded contracts, requiring no more than that they accomplish the research and obey the relevant laws while doing so. In my opinion, the federally controlled laboratories should be phased out and their assets should be sold.

RESEARCH, REGULATION, AND THE UNKNOWN

Our discussion so far has focused principally upon research whose purpose is national defense. Much of what we have said, however, would apply equally well to research conducted for other purposes. For example, police agencies might choose to support research aimed at computerizing fingerprint detection or research aimed at developing psychological profiles of terrorists. These would be legitimate, as they both support the valid purposes of government, and both could be done wisely. Beyond these defense and police functions, government might need to support research for other reasons related to the other valid purposes of government discussed above in the second section. It is unfortunate that we do not have space to explore this as a general subject here, because the specific subject of regulation is of great practical importance.

It is unclear to me whether regulation is a necessary component of a classical-liberal government. Many classical liberals advocate the use of the courts to deal with all circumstances in which citizens can harm one another. Some would allow statutory law to play a role as well. However, what is now known as the environment is in many respects a natural commons. Problems such as air pollution, ocean pollution, and harmful-insect propagation do harm citizens and their property but do not obviously lend themselves to strict judicial solutions. Much work in political science is needed here. It is very unclear in what sense and to what extent a citizen has a right to clean air or a right to the present climate. It is even more uncertain whether and to what extent regulation will be needed to secure these rights.

It is very clear, however, that governments will be using regulatory mechanisms for the foreseeable future. Because of this, work is also needed on how regulation can be done wisely. Libertarian think tanks are doing some of this today, for specific applications. We can hope that someone is undertaking (or soon will) a more general analysis of how to regulate wisely, which is outside our scope here. A related and interesting question is whether the government ever should engage in regulation of some activity when research about its effects has not yet produced certain results. The answer here might be yes, if the plausible or probable consequences are large enough. In any event, governments today routinely implement regulations under such uncertain circumstances. As a result, the implications of this situation for research deserve our attention.

In considering the interplay of research and regulation, it is crucial to note that the mindset and the incentives for these two activities are completely different. The mindset of the researcher is to discover how reality works. This includes the need to identify and focus on areas of uncertainty, and to continually question what seems to be well known. The mindset of the regulator is to devise a practical means of restricting human action, to achieve some simply stated goal. The goal often reflects scientific work, but cannot respect scientific uncertainty. For example, the decision to regulate often requires that an acceptable concentration or rate of emission of some pollutant must be established. For two reasons, the established limit very often will be based on uncertain science. First, the question, "how small is small enough?" is often difficult for science to answer. Second, the political process produces a drive to act NOW, in response to whatever threat is perceived. The incentive for the regulator is to specify some standard and to implement a regulation. Politically, it matters much less whether the regulation is soundly based.

There are two implications for research. First, research will very often be needed and important *after* a regulation is enacted. The continued development of scientific knowledge will often show that the decisions made when enacting the regulation should be revisited. The science will sometimes show that the regulation is very wrong. Yet there has been a clear trend for government agencies to cease supporting research as soon as a regulation is in place. The Department of Transportation stopped its support of stratospheric ozone research as soon as regulations were in place regarding chlorofluorocarbons. The interagency acid rain program was terminated as soon as emission controls, intended as a response to acid rain, were in place. After the EPA proposed regulating hydrocarbons to control ground-based ozone, research at Georgia Tech showed that this is often impossible. The conclusion to draw here is that *regulation requires research*. The government should assure that there is continued development of the science in any area in which it chooses to regulate. This may mean that the government should fund such research, if it will not be accomplished by other means. It should do so in order to assure that the regulation is achieving its stated goal of protecting the citizens in some way and to assure that the freedom of action of the citizens is not being restricted unnecessarily.

Second, the incentive of the regulator is to oppose such further research. It makes the regulator look bad when the regulation needs to be changed. Dealing with such issues also distracts the regulator from the process of developing other needed regulations. An example from California illustrates the incompatibility of regulation and research. California gave the problems of both regulating air pollution and supporting research to its California Air Resources Board. The board established an advisory committee of scientists to advise it on research directions, and then proceeded to blatantly ignore the advice of the committee. The committee felt that its advice was not taken seriously, and disbanded in protest.

One can conclude from the above that *one must separate research and regulation*. Only this can provide a system of checks and balances that will dampen the tendency to regulate excessively and unreasonably. If one establishes a regulatory agency, one should also establish independent research within some other agency. The task of the independent research agency should be to advance the science related to the regulations without regard for the political consequences. (This will be politically practical only if the research agency supports research by others, through grants and contracts, but does no research itself.) The United States combines research and regulation in several cases, notably including the Environmental Protection Agency and the Food and Drug Administration. Based on the above discussion, this is a colossal mistake directly responsible for many of the bad regulator decisions that one hears about from these agencies. One does not hear about many other cases, in which further research, never undertaken, would have shown that some regulation ought to be amended or scrapped.

CONCLUSIONS

We have seen that the General Welfare clause in the U.S. Constitution is the underlying principle that justifies government support of research today. Some of this research is done wisely; some of it is not. However, once we reject the General Welfare clause, and seek a government based on the defense of individual rights, we still find that the government should support research. Unfortunately, it does not prove trivial to place limits on which research should be supported, and we end with the problem of needing to proceed wisely in the face of complex trade-offs. A historical review suggests that much of the research supported during the twentieth century was appropriate for defense reasons, although only some of the methods of support were wise. In particular, both federal labs with Civil Service employees and Federally Funded Research and Development Centers ought to be phased out and replaced by private alternatives. We have also seen that regulation and research are conflicting missions. In consequence, a government that would regulate wisely must separate regulation and research.[28]

28. The author acknowledges useful discussions with David Kelley, Joyce Penner, Patrick Stephens, and Roger Donway.

CHAPTER THREE

Federal Support of Research and Development in Science and Engineering

Michael W. Blasgen

SUMMARY

Federal funds pay for about a third of all research and development performed in the United States.[1] The effect of these funds is pervasive. As we will discuss, this funding is either based on principle, or politics, or effectiveness. If we consider only the role of government as a funder of research and development, then principles are pretty few and far between. Politically, there is huge support both by politicians and the public. Finally, on effectiveness we will see that it's a checkered history, as one would expect in a game that is played by Washington rules, further complicated by the nature of research and development work: it's like betting on long shots.

This essay discusses the history and effects of years of federal funding in the physical sciences and engineering. In some ways the effect of fed funding on medical and biological sciences is

1. National Science Board, *Science and Engineering Indicators—2000* (Arlington, Va.: National Science Foundation, 2000). *http://www.nsf.gov/ sbe/srs/seind00/frames.htm.*

more dramatic, but my background leads me to focus more narrowly, specifically on information technology (IT).

What have the feds funded in IT? Consider this: there is a list of the five hundred most powerful computers in the world.[2] Of the top one hundred, two are owned by private industry to solve problems, the ninety-eight others are owned by government (including non-U.S.) entities or contractors, and the supercomputer vendors themselves. The bulk are in the United States. This is deliberate policy of the National Science Foundation (NSF), Department of Energy (DoE), and the Department of Defense (DoD): to build in the United States the largest infrastructure of supercomputers to keep ahead of our military and economic competitors. Is this a good thing? On the face of it, it sounds like good logic: the computer industry has fueled the recent economic growth in the United States (the Internet, the Web, etc.) and we need to keep priming that pump. If there is a flaw, it is that the supercomputer segment has not grown, in fact most vendors have failed or been acquired in unfavorable circumstances. Control Data, Cray Research, Cray Computer, Thinking Machines, and many others have been eliminated as organizations as mainstream computing environments (based typically on PC technology) take over. Yet the development of the PC was not funded by the feds. In fact, of the technologies that were needed to enable the PC revolution, only one, networking, was created and supported in any serious way by government funding.

Certainly the Internet and Web owe a great deal to government funding. Could this have happened without the government funding? We won't ever know the answer to that, but the fact that proponents regularly point out this one success

2. TOP500 Supercomputer Sites. *http://www.top500.org/*.

story after thirty years of federal funding suggests that the funding has not been as effective as one might hope.

INTRODUCTION AND BACKGROUND

In the United States, federal funds provided for research and development dominate the research and development community. Approximately one-third of all research and development conducted in the United States is paid for by federal taxpayers.[3] The federal government's research and development spending is $62 billion, not counting tax expenditures, out of a total spending of about $200 billion in the United States. Research and development includes a lot—making minor design modifications on a tank or ship design is research and development, but the federal role in more basic research and development is even more dominant. The support for basic research by the feds is between $21 and $29 billion dollars per year (it depends on how you count), an amount that has doubled in the past ten years, and that represents more than half of all of this type of research. In comparison, this third or half is far more than the role of the federal government in the overall economy: federal spending is 18.2 percent of Gross Domestic Product.[4]

Is this dominant federal role appropriate or justifiable? Research and development is not mentioned in the Constitution or any of its amendments and would appear to be a constitutionally unjustified intervention in the economy, yet research and development funding is so deeply embedded in role of the

3. National Science Board, *Science and Engineering Indicators—2000*.
4. *FY2002 Economic Outlook*, Executive Office of the President. *http://w3.access.gpo.gov/usbudget/fy2002/pdf/economic.pdf.*

central government that it is listed by the president in his economic reports as one of the sixteen basic roles of government.[5]

The justifiability of research and development support by the U.S. government can be discussed on principle, or on political realities, or on effectiveness.

POLITICS

Politics can be quickly dispensed with: there is NO political debate on the funding of research and development. The two political parties that dominate U.S. politics compete to see which one can cause the biggest increase in research and development outlays. George W. Bush is raising NIH's budget by a substantial fraction over the proposed Clinton budget. Senate votes on budget for the National Science Foundation can be 95 to 5, or even when bundled with HUD's budget will attract a vote of 87–8.[6] I recall that near the end of his term President Carter nominated a new head of the National Science Foundation, and the Republicans blocked the nomination so that Reagan could nominate the same person and get credit.

Politicians are reacting to public support. In 1999 a survey found that 82 percent of those queried agreed that "even if it brings no immediate benefits, scientific research that advances the frontiers of knowledge is necessary and should be supported by the Federal Government."[7] I can think of no area of government where the expenditures are so great and the controversy so small. Why no controversy? Perhaps because scientists are held in high esteem by the voters. Perhaps because

5. *FY2002 Economic Outlook*, Executive Office of the President.
6. "On Thursday, October 12, 2000 the Senate passed H.R. 4635—the final Senate VA-HUD and Independent Agencies appropriations legislation for FY 2001. The vote was 87–8." From NSF home page *http://www.nsf.gov/*.
7. National Science Board, *Science and Engineering Indicators—2000*.

the funding really has worked—that without it we would all have a much lower standard of living. Perhaps because we as a nation want to win as many Nobel prizes as possible and will spend whatever it takes. Perhaps because we want the highest technology for our armed forces, and realize the amount of research and development that actually takes. Perhaps it's the health research driven by our fears of death and disease.

Clearly, based on the political realities, there is adequate basis for federal funding of research and development.

PRINCIPLE

As to principle, there is no shortage of descriptions of the benefits of government-funded research and development, usually sourced from the government itself, or from agencies supported by government. Government documents and congressional testimony explain why we spend so much money on research and development. It's "necessary to the vitality of the nation" or "contributes to economic growth and productivity." The National Science Board, in its biannual report *Science and Engineering Indicators*,[8] states, "Research and development is a vital and necessary step leading to improved products and services."

The president's *Economic Outlook 2002*[9] says research and development outlays "help the Nation compete in the global economy and improve our quality of life." The president's report goes on: "In the last fifty years, developments in science and technology have generated at least half of the Nation's productivity growth, creating millions of high-skill, high-wage jobs." The basis for adopting this role of government is com-

8. National Science Board, *Science and Engineering Indicators—2000*.
9. *FY2002 Economic Outlook*, Executive Office of the President.

paratively recent. It has to do with the changes the United States caused by World War II.

Prior to World War II, what little research and development funding the U.S. government did addressed specific problems such as agricultural blights, human disease, and calculation of weapon ballistics. Then the world descended into war, and it was a different kind of war, a wizard war. Technical breakthroughs in radar, sonar, magnetically fused depth charges, and code-breaking, to name a few World War II-inspired technologies, had never been used before, yet changed the outcome of the war. The key developments were carried out by "wizards," technically trained people, many of them physicists, who understood these advanced technologies. In the early years of the war England led the development of radar (an acronym for radio detection and ranging) and pioneered electronic code-breaking. In the thirties the United States had been rather far behind in understanding and exploiting science in general. German universities that subscribed to U.S. physics publications arranged to have an entire year's volumes mailed in a single package to save on postage. That meant they would see the articles as much as a year late, but there was no urgency to read U.S. publications because it was so unlikely that any important science would be done in the United States.

By the end of the war this had changed. The War Department had mobilized the best minds in the United States, who were able to take advantage of government money and, unlike the British, were at little risk of being bombed. The Manhattan project (using the physics developed earlier by Europeans) successfully designed, built, and detonated two designs for atomic bombs, bringing the war to an end. The Office of Scientific Research and Development, headed by Vannevar Bush, was formed within the Department of War. Once the end of the war was in sight, FDR formally requested Bush to suggest how to continue the contributions in peacetime. FDR wrote, in part:

> Dear Dr. Bush: The Office of Scientific Research and Development, of which you are the Director, represents a unique experiment of team-work and cooperation in coordinating scientific research and in applying existing scientific knowledge to the solution of the technical problems paramount in war. Its work has been conducted in the utmost secrecy and carried on without public recognition of any kind; but its tangible results can be found in the communiques coming in from the battlefronts all over the world. Some day the full story of its achievements can be told. There is, however, no reason why the lessons to be found in this experiment cannot be profitably employed in times of peace. The information, the techniques, and the research experience developed by the Office of Scientific Research and Development and by the thousands of scientists in the universities and in private industry, should be used in the days of peace ahead for the improvement of the national health, the creation of new enterprises bringing new jobs, and the betterment of the national standard of living.[10]

I'm sure that this letter did not surprise Dr. Bush, who in turn wrote a detailed document entitled "Science—The Endless Frontier" describing how the postwar government scientific enterprise should be constructed. The report proposed the creation of the National Research Foundation, an organization that became the National Science Foundation (not quite right, since the first research issue discussed in the report is

10. Vannevar Bush, *Science—The Endless Frontier: A Report to the President, Director of the Office of Scientific Research and Development* (Washington, D.C.: United States Government Printing Office, June 1945). *http://www1.umn.edu/scitech/Vbush1945.html.*

"the war against disease" which ultimately became the basis for the National Institutes of Health). The report also directly addressed the issue of whether or not the support for science was a proper function of government. On that topic, the report states that "science IS a proper concern of government" and goes on to say:

> It has been basic United States policy that Government should foster the opening of new frontiers. It opened the seas to clipper ships and furnished land for pioneers. Although these frontiers have more or less disappeared, the frontier of science remains. It is in keeping with the American tradition—one which has made the United States great—that new frontiers shall be made accessible for development by all American citizens. Moreover, since health, well-being, and security are proper concerns of Government, scientific progress is, and must be, of vital interest to Government. Without scientific progress the national health would deteriorate; without scientific progress we could not hope for improvement in our standard of living or for an increased number of jobs for our citizens; and without scientific progress we could not have maintained our liberties against tyranny.[11]

The principle on which today we spend $62 billion annually is the opening of new frontiers. I think Lewis and Clark would be amazed.

By the Eisenhower administration, the science-government alliance was in full flower. There was a science adviser to the president. Eisenhower realized that the most powerful weapon in the world, the weapon that had ended World War II, had been designed and built using science so advanced that almost no American had even a small understanding of the physics underlying a fission bomb, much less a fusion bomb. The military was ever more dependent on technology—jets were re-

11. Vannevar Bush, *Science—The Endless Frontier*.

placing propeller planes, radar was critical, the English code-breakers were supplanted by the (now vast and powerful) National Security Agency—and the support of research and development seemed essential. The military became such a force in the United States that Eisenhower, a former army general, said that we needed to be careful of the new military-industrial complex. Since then, Congress and the administration have supported increased research and development budgets. Today, as stated earlier, the federal government spends $62 billion annually to open new frontiers.

EFFECTIVENESS

Whether or not research and development spending is principled or politically necessary, the best practical reason for funding it is that it is more effectively done and used by the state. The remainder of this chapter will discuss this issue.

Is it effective? Recall the sentence, quoted above, from the President's *Economic Outlook 2002*: "In the last fifty years, developments in science and technology have generated at least half of the Nation's productivity growth, creating millions of high-skill, high-wage jobs." Actually the paragraph goes on: "Federal Government support for science and technology has helped put Americans on the Moon, harnessed the atom, tracked weather patterns and earthquake faults, and deciphered the chemistry of life."[12]

What's going on there? It's a pretty carefully worded sentence. Does it say that government support of research and development has helped productivity and created jobs? My guess is that this is what the author hoped you would conclude, but it doesn't actually SAY that. Instead, the author first ob-

12. *FY2002 Economic Outlook*, Executive Office of the President.

serves that developments in science and technology have helped productivity, and that's probably true. But what about federal support? The author says that this allowed us to go to the moon, build nuclear weapons, and so on. But those accomplishments have probably NOT contributed to productivity and high-wage jobs. Thus the paragraph does NOT say that federal support has helped productivity, although most readers would probably conclude the opposite. Unfortunately that's true of a lot of research and development funding: we say it helps the economy or competitiveness or vitality or productivity, but most of it doesn't.

In truth, more than half of research and development spending is done in just two areas: defense and health. Defense is a constitutionally authorized area, and health research has widespread political support because everyone is affected by issues of health and longevity. Perhaps that should end the discussion, but, on the face of it, spending in neither area makes the United States more competitive (except militarily), nor more productive, nor does it improve the standard of living. Since this is the bulk of spending, and we are all wrapped up in our rhetoric about productivity, we argue that defense research and development will provide technology that benefits U.S. industry. In particular, the argument advanced for most research and development is that the work funded by the feds will trickle down into the economy. The only problem with that argument is that there are very few examples where that has happened. As the *Wall Street Journal* stated recently: "Defense industry executives go all the way back to the passenger jetliner to name a military advance that found mainstream popularity."[13]

To discuss the question of effectiveness in more detail, this

13. *Wall Street Journal*, April 25, 2001, A1.

chapter now explores the role of federal funding in just one area: information technology. Computers are important to the economy, and federal research and development money was used, so it's a good case. In fact, it's the one most likely to be rolled out in defense of federal involvement in research and development. And it is also the area with which I have the most personal familiarity, as I am a computer scientist.

SOMETHING ABOUT ME

Before discussing the involvement of the U.S. funding of computers and networking, let me provide some personal background. My education and employment in the computer industry cover the period 1960 to 2001. I have an undergraduate engineering degree from Harvey Mudd College, an electrical engineering master's degree from Caltech, and a Ph.D. from the University of California at Berkeley. My Ph.D. is in electrical engineering and computer science (EECS). That name reflects not only my history, but the history of many of the folks who jumped on the computer bandwagon in the sixties and seventies. When I enrolled at Berkeley in the 1960s, I joined the Electrical Engineering Department; however, it was clear to everyone that the important application of many of the principles of electrical engineering would be found in digital computers. (Just for the record, there were analog computers—the term digital computer is not redundant.) Thus I and many of my colleagues became involved with computers, and the name of the department itself had changed by the time I completed my degree.

By the time I was ready to get a job after receiving a Ph.D., I had worked at six organizations in technical positions: a Navy laboratory, System Development Corporation (a spin-off of Rand Corporation to run missile defense and air traffic control), Hewlett-Packard, Bissett Berman (a company with a prime NASA contract to do navigation on the Apollo moon missions), Stanford Research Institute (now called SRI), and University of California, Berkeley. All but one of these organizations was carrying out research and development using federal funds and under federal contract.

My work has always had to do with computers, from relatively early computers from the late fifties to modern machines. Let's look at the role of the feds in the development of the computer industry.

EARLY HISTORY OF COMPUTERS

Computers were developed to calculate answers to numeric problems. (This is distinct from punched-card equipment, mostly made by IBM, that was used by the census and by insurance companies.) Early adopters of computers had numeric questions: where was Jupiter going to be next week? This led IBM to support astronomers at Columbia in the 1940s. More commonly, computers were needed to figure out where a shell fired from a battleship was going to land. This application drove most of the early computer designs, mostly paid for with Department of War money. A few brilliant scientists and engineers realized that a computer could compute more than just numbers, and developed computers that could, for example, break German codes. The ballistics and code-breaking applications were peculiar to defense needs, and thus the early support for these computers came from the defense establishment. As Encarta states, "At first, only the government used computers." There are few examples of early government support leading to such an important and broad-sweeping industry.

HOW COMPUTERS EVOLVED

Soon after computers were developed by universities and government labs, private companies jumped into the business. The most famous of these was Univac, a name that was almost synonymous with computers. (The Univac name and business were later purchased by Remington Rand.) These were companies that focused on the applications the government wanted and received substantial federal support for their research and development to develop products. In 1951, to pick a date when the outcome was quite in doubt, it appeared that the "giant brains" would be used only by the government, working on very large problems, because the machines were so expensive and massive that only a project like the census or the military could use them. Computers were certainly thought of as a new thing but no one predicted what would come next.

What came next was a rapid conversion of the electronic data processing business from punched cards to computers that in turn used punched cards. It turned out that expertise in punched cards (originally developed by Herman Hollerith to help the census of 1890) was vital to success in the computer business, and as IBM fortunes rose, the fortunes of Univac and the other computer companies declined. While it's probably an exaggeration to say this, Univac declined while focusing on the special needs of government and IBM thrived by satisfying business needs.

BACK TO ME: CHOOSING AN EMPLOYER

By the sixties, the role of federal funding in the evolution of computers had declined, but they were still the dominant supplier of research funding in computer science. Going back to my personal experience, by 1969 I had worked at six companies, and five of them were doing computer research funded by the feds. The one exception was Hewlett-Packard, where I and others in the laboratory were designing instruments like signal generators and oscilloscopes using HP's internally generated funds. As a practical matter, someone with my training was most likely to wind up working for a government contractor or a university, but also for organizations like SRI.

My experience with contractors caused me to try to avoid federal contracts. There were just too many problems associated with federal funding: chasing the contract itself was a huge but largely nontechnical project that consumed lots of time and energy, and it was anticlimactic when the contract was acquired. The scope of work seemed to be controlled by bureaucrats located far from the lab. The focus was too much on defense and space, areas that seemed to me to have little hope of ever seeing application in the real economy. Even if the project did not appear to be a defense project, it always had to be justified as important to the Army or whatever. A funding agency might say, "We need natural language understanding by computers so that the captain can control the ship." While today there are many privately funded jobs available in the IT (Information Technology) industry for skilled engineers and computer scientists, that was not true thirty years ago. Even today, most researchers who are interested in fundamental problems in computing work in the major universities that are large federal contractors. Sixty percent of all research and development conducted by universities is paid for by the feds.[14]

I decided that I didn't like the life of a government contractor, but in 1969 if you didn't want to work for the government,

14. National Science Board, *Science and Engineering Indicators—2000.*

where did you go? I joined IBM, which had a research laboratory in Yorktown Heights, New York, that was generously funded from internally generated funds. At the time there was only one other lab like it—Bell Labs in Murray Hill, New Jersey, funded from internal AT&T funds. As it turned out, the decision to avoid government contracts was wonderful— it led to a long and satisfying IBM career. It's a wonderful illustration of being able to act on my view that government's role in the economy should be minimal. At least I could take myself out of direct dependence on government funding.

SIGNIFICANCE OF
INFORMATION TECHNOLOGIES

There is little question of the importance of the computer industry in the economic life of our nation and the world. *Science and Engineering Indicators—2000* starts chapter 9 with the statement that IT industry growth is responsible for 29 percent of growth in national income in 1998, and declining prices of IT equipment reduced inflation.[15] While the chapter does not claim a connection between research and development funding and the importance of the IT industry, the general theme of the document certainly implies it.

Where is the government spending money now? Where did it NOT invest? Let's look first at an area that has received special treatment from the government, and a large amount of funding: supercomputers.

15. Ibid.

SUPERCOMPUTERS

Government has been supporting the construction of the biggest computers since the beginning of computing. I guess it's like bombs and airplanes, if we can build a bigger or faster one, let's do it.[16] Even today, the federal government pays for the largest computers. Of the largest and most expensive computers in the world, all are owned by government labs with only one exception: a machine at Charles Schwab.[17] How is it that government owns the most expensive computers in the world? They don't own the best cars, the best tractors, the best steel mills, the best stereos; what is it that has them own the best computers? To answer this we can look at the uses to which these machines are being put: all but a few of the top twenty machines are applied to weather prediction, nuclear weapon design, code-breaking, and high-energy physics. Let's consider these four applications.

Weather? It is a hard problem, is helped by faster computers, and is the more or less exclusive realm of the government because they give away the information. The price they ask makes it difficult for a competing private institution. Code-breaking and signals analysis? Again, only the government imagines that there is that much value in listening to the electronic signals of economic competitors and rogue nations. High-energy physics? Again, no economic value—this is really

16. Spring Joint Computer Conference, *AFIPS Proceedings* 36 (1970): 543–49.
17. The Top500 organization cannot track ALL computers. It tracks the ones it knows about. There is probably a bias in their choice of computers—the ones listed by them tend to be computers used in scientific computation. For example, Google (*www.google.com*) or Yahoo (*www.yahoo.com*) may own computers that in the aggregate are large enough to qualify for Top500 status, but they are not listed. *http://www.top500.org*.

research, and the prestige of the nation is thought to be on the line. We get a disproportionate number of Nobel prizes in physics, and we want to keep getting them. Nuclear weapons? No one wants them except governments. To test this, let's look at the list of the top one hundred machines. Ninety-four are owned by government labs, government-funded centers, or by the vendors themselves for testing. Six are owned by industry.

Each of these multimillion-dollar supercomputers is applied to a problem that only a government would care about, but why do they use supercomputers? The U.S. Congress has little problem justifying the billions spent each year on acquiring and operating these machines, and yet private industry spends almost nothing on these monster machines. That's because they are much too expensive for the return. Here government has a Manhattan-project attitude: if there's a way to spend more money to do a better job, then spend it. In contrast, commercial enterprise looks not at effectiveness, but at cost effectiveness. For most problems (the Charles Schwab machine being a possible exception) supercomputers are not the best solution.

What is the best solution? The personal computer. There must be some good reason why there are millions sold each year, in fact, for the vast majority of IT issues, personal computers and the servers (mostly based on the same technology as the PC itself) that support them are the technology of choice, so what did the government have to do with the productivity-enhancing IT systems that we all use today? The key inventions that led to the small high-performance systems that we have today can be placed in seven categories. I have listed one or two of the companies that are influential in this market.

overall design	IBM, Apple
applications	Microsoft

middleware (e.g., database)	Oracle, Microsoft
system software	Microsoft
networking	Cisco, 3Com
storage (e.g., disk drives)	IBM, Seagate
microprocessors and DRAM	Intel

What was the role of government funding in this? We will explore each in turn, but the bottom line is that one area, networking, of these seven areas has been led by fed-funded research and development. In the other six areas, the federal funding has not been a crucial enabler.

SYSTEM ARCHITECTURE

The stories of the invention of the Apple PC and the IBM PC have been described in many places. Suffice it to say that government money played no role in either design effort.

APPLICATIONS

Whether we consider personal productivity applications like Microsoft's spreadsheet and word processing or if we look at SAP's enterprise applications, there has been no significant government funding involved.

MIDDLEWARE

For the most part, federal funding played little role in the development of companies like SAS Institute, Oracle, and others. Yes, the very first version of Oracle's system was developed for the CIA under contract, and yes, some key pieces of middleware associated with networking (in particular the web browser that led directly to Netscape) were funded by the feds, but beyond this there are few examples of federal funding of research and development that contributed to today's U.S. dominance of middleware.

The federal funding of software has played a small but important role in the evolution of software. One example: engineers at AT&T Bell Labs using AT&T's money had developed as a research project an operating system called Unix. Because AT&T was prevented by law from entering the computer business, it did not sell Unix directly but instead licensed various nonprofit organizations to use it. It was the original open source product, at least in the sense that the organizations that licensed it received source as well as object.

The Department of Defense funded the support and enhancement of the original Unix in an effort to rationalize the computing environments used by the many Department of Defense research contractors. This effort was known as Berkeley Unix, or BSD (Berkeley software distribution). This has certainly been effective in laying the groundwork for several important developments. Sun Microsystems was founded as a Unix workstation company, using BSD and in fact hiring the key programmers from the University of California, Berkeley. The whole Open Systems movement revolves around Unix; the Gnu software complex that includes the Gnu compilers again revolves around Unix. In fact, Gnu itself is an interesting self-referential acronym—Gnu stands for "Gnu is Not Unix." However, Unix was not able to develop into the foundation of most computing. That role was taken by Microsoft. Originally Microsoft took their operating system inspiration from CPM but later was influenced by Unix. Of course Unix itself was influenced by work that preceded it.

Net: while U.S. spending on system software had some impact, the major impact of Microsoft's system software did not depend in any important way on previous federal support of system software research.

NETWORKING

In this area there has been a major contribution by research and development funded by the government. This is worth a whole section of this article.

STORAGE

The key development in this whole area is the disk drive originally developed by IBM using internal funds. IBM's ideas were later exploited by Seagate, who pushed the technology into smaller packages. The feds have supported research institutions at Carnegie-Mellon University and others on pushing the envelope of magnetic recording, but this has not been critical to the amazing advances of magnetic recording being made by the privately funded disk-drive industry.

Department of Defense did support a small research project at Berkeley that led to RAID (reliable array of inexpensive disks), and the RAID work has been critical to the evolution of storage systems.

DRAM

Originally modern DRAM (dynamic random access memory) was invented by and developed for industry. An engineer at IBM invented the concept of DRAM, and IBM first introduced DRAM (also known as solid-state memory) in its computers to be used, as usual, for a company that called itself a business machine company, for customers like insurance companies. Later, again in the late 80s, the feds played a role in DRAM. It's the same story as the role in silicon: the United States was concerned that all DRAM manufacturing would move to Asia. Since DRAM is a key technology, the United States offered what amounted to subsidies to companies manufacturing DRAM in the United States. Today most DRAM is manufactured offshore, and most of those companies do not make much money doing it.

MICROPROCESSORS

The evolution of microprocessors, from the early Intel chips that were embedded in Japanese calculators to the latest Pentium IV, was driven primarily by industrial requirements. The government did not play a major role. This story is not dissimilar to the technology story: for the most part the Department of Energy and the Department of Defense thought they wanted the highest possible performance. They were led thus to supercomputers, and microprocessors were perceived to be too weak to satisfy serious computing needs. NASA needed compact machines for onboard computing, but man was already back from the moon before the microprocessor was developed.

The primary conclusion to be drawn from this summary of the key technology underlying modern computing is that the feds support of information technology research and development did not play a decisive role in forming the industry, with one notable exception: the Internet. It is therefore necessary to look into this story.

THE FEDERAL ROLE IN
CREATING THE INTERNET, THE WEB,
AND ALL THE NEW NEW THINGS

I recall clearly the day I went to my student office at Berkeley and observed a big moving van parked in front of the building. A new project, the ARPANET, funded by the Advanced Research Projects Agency of the Department of Defense, had recruited one of the faculty stars from Berkeley, and the van was moving his office contents back East. That piqued my interest, and I carefully read all the documents that were produced over the next ten years. The first paper I read was by Larry Roberts on the architecture and motivation for the network. The title provided the primary justification: "resource sharing."[18] This was an interesting idea. In the sixties and seventies, every research and development center—mostly universities—wanted a big computer. ARPA's funds would be quickly depleted if every contractor was provided with the biggest computer available. Recall that these computers could cover more than ten thousand square feet, requiring enough air-conditioning to run a city. They cost millions and were a big prestige symbol. When IBM built a new research lab in San Jose, California, the design concept was a large machine room with many offices overlooking it from above; as though the best view California offered was that of a big computer. The idea behind the ARPANET was to offer to every ARPA contractor access to all the big computers, independent of geography. That way researchers in Wisconsin could get access to the big IBM computer at UCLA, and ARPA wouldn't have to buy one for Wisconsin, or so they hoped. It didn't really work

18. Lawrence Roberts and Barry Wessler, "Computer Network Development to Achieve Resource Sharing."

out that way. Instead, the ARPANET became the Internet, and certainly became the most important development in information technology. Not only did the ARPANET lead to the Internet, but it also drove competing designs away. While this might be considered a bad thing—"What were the feds doing funding stuff that had the effect of destroying all that privately funded work on networking"—in fact, it was the right thing technically. I saw this from the IBM perspective.

ARPA had encouraged (and sometimes required) widespread distribution of the information on the network design. The protocols, the packet design, and the software were all widely examined and understood. In contrast, IBM's networking architecture was different not only in detail, but in the openness of the design. IBM's computer networking was called System Network Architecture (SNA) and was a proprietary session-oriented approach that was very different from the ARPANET direction. Further, IBM and certain other major research and development centers were isolated from the ARPANET itself because the feds did not charge for the use of ARPANET and therefore you had to be a major ARPA contractor to get access to the network. I recall a meeting on this in the mid-seventies. IBM had hired some professors from MIT who were intimately familiar with the ARPANET to come and provide advice on the evolution of SNA. The MIT professors wrote a report that said, in effect, "we don't really understand what IBM is doing so we can't really help you." Whoa! The consultants said that IBM should spend more time thinking about the protocols used in the ARPANET. As it turned out, they were right and SNA slowly died.

The ARPANET/Internet/Web is a wonderful development—like the personal computer it has become ubiquitous—and has scaled a degree never imagined by its designers and developers. There is no question that the federal funding of the initial ARPANET was responsible in a major way for the success of this major innovation. We can't know how it would have turned out without the ARPA money; but we do know that the Internet today is a direct descendant of the work done in the late sixties by a bunch of very talented computer scientists working under the leadership of a couple of guys in the Department of Defense. Without question, this is an accomplishment. But even in a success story we can see the pitfalls of government funding of research and development.

A MAJOR PITFALL: SATISFYING THE FEDERAL RESEARCH AND DEVELOPMENT CUSTOMER

Effectiveness is a key metric of judging federal research and development funding, absent a sound principle on which to base the spending. The experience of World War II led directly to the major involvement of the government in research and development. The feds have supported many areas of computer research and development, and regularly draw attention to the huge success of the Internet. Other success stories are fewer and farther between. Recall the list of companies involved in the early evolution of the modern computing environment? To repeat:

overall design	IBM, Apple
applications	Microsoft
middleware (e.g., database)	Oracle, Microsoft
system software	Microsoft
networking	Cisco, 3Com
storage (e.g., disk drives)	IBM, Seagate
microprocessors and DRAM	Intel

Which one of these received major fed funding? None, I

believe. Where's Bolt Beranek and Newman, a major contractor on the ARPANET? Where are the NASA and Department of Defense contractors who received vast sums to develop computing environments? You rarely hear about the contributions of General Dynamics or Lockheed to modern computing. My guess is that the companies that depended on federal funding looked to satisfy the customer (as they should) and that meant satisfying the funding agencies. It's a rare case when satisfying a funding agency leads to commercial success in the competitive world marketplace.

Consider the companies below, some of which got lots of research and development money from the government, and others got comparatively little. Which have done better over the years?

- Didn't get money: Apple, Cisco, Intel, Microsoft, IBM
- Got lots of money: North American Aviation, General Dynamics, Bolt Beranek and Newman, Lockheed Martin, Computer Science Corporation

A MAJOR PITFALL: LACK OF DIVERSITY

Think of Adam Smith's invisible hand. It picks winners based on marketplace competition. How does that work in fed funding? Well, there's only one federal funder of research and development. No invisible hand is going to come around and knock off that funder and replace it with another. Thus we have a monopoly regulated only by the processes of fed funding. Those processes include getting good people to propose project areas and oversee the review processes, peer review and competitive bidding, along with the usual pork-barrel rules.

There is no question that NSF and ARPA get good people

to spend a portion of their careers helping to manage the research and development spending. They are certainly well-intentioned, but face no test of market effectiveness. Peer review tends to ensure the scientific worthiness of a proposal, and is effective in weeding out technically worthless ideas. It does not, however, help to FIND ideas. In fact, there is a large "piling-on" effect in fed funding of research and development. The project manager decides that a certain area should be funded. For example, at one point the feds made it clear that they wanted to fund research and development in software engineering tools and object oriented programming, and the proposals poured in. The ideas followed the dollars, not the other way around.

Following the fashion is also a defect of privately funded research and development, but not nearly as serious. Much more diversity arises in privately funded research and development. Of course the federal funding agencies try to find out how effective they are by counting papers published, or encouraging industrial matching money (but this is hard—see below), or by looking for commercial applications of fed-funded research and development, but it is fundamentally impossible to support the kind of research and development diversity that arises in a free market.

Mixing federal and private money to fund a specific research and development project has always been difficult. Initially places like NSF did not want any confusion about how their money was being used, and insisted that any results from the research and development paid for by NSF be placed in the public domain. In the past ten years the funding agencies have softened this stance and are trying to encourage joint funding, but it is still difficult. Basically the fed funding agencies are very sensitive to the criticism that they are giving public monies away to private companies. It's pretty much OK to give public monies to colleges and government agencies, but not private companies, so they try to get agreement that the results will be openly published and that any patents will be made available to others. This conflicts with the needs of private industry, which usually spends the research and development money to gain a competitive advantage.

A MAJOR PITFALL:

FUNDING IS A MONOPOLY

Since sports metaphors appeal to some readers, imagine the following variant on baseball. You are given the bat. You stand at the plate, and try to hit the ball. You can swing as often as you want since there are no strikes and no balls, and you can just wait for your pitch and try to hit it. There is no penalty for swings, or fouls, or hitting poorly. After some long period of time, we count up the good hits and say, "Amazing, you hit forty-seven singles and fifteen doubles. Great!" When there's only one hitter, and no limit to your swings, you are BOUND to get hits. Thus it is with research and development funding. Eventually something will work out. Then, like the Internet, it is touted as a major accomplishment (which it is, after all). But is this the way to play the game?

Like our imaginary batter, there is ultimately one organization responsible for this government support of research and development, the government itself. It decides, and given the lack of public debate, the staffs who run the funding agencies decide, who wins and loses. The marketplace of ideas, science, and progress itself, is controlled by bureaucrats, however competent, in Washington.

CONCLUSION

We lack a guiding principle for federal funding of research and development. Because there is so much public support and political support for research and development spending, the public debate that normally deals with political decisions is largely missing, so we are left with the argument that it works. We've seen that it can work. The feds funded the key work that led to the Internet and Web (a good thing) and has led to the dominance of government-owned supercomputers (a bad thing?).

The U.S. government has spent a few hundred billion dollars of taxpayers' money on research and development in the past two decades. The politicians are happy, citizens like it, and staffs at NSF, ARPA, and similar funding agencies are competent. The funding has produced some home runs like the Internet. We'll always have a need for a certain level of government-funded research and development given our desire to have the most powerful army in the world. Maybe there's no issue.

But there are issues because there is no principle to justify or defend the spending. "Vitality of the nation" is simply not sufficient. That's the way you justify supporting the symphony or ballet, not $62 billion in spending each year. There is no substantive political debate. Finally, in spite of many efforts to track it, there is really no way to establish that the money is effectively spent. It is neither enough to count Nobel prizes (that's the first table in Science and Engineering Indicators) nor it is sufficient to cite the success of the Internet, however great that accomplishment is.

We've described three pitfalls in the funding system:

- There is a virtual monopoly on running the funding system.

- The monopoly creates a lack of diversity in ideas.

- The folks who play the grant-contractor game the best are often least likely to succeed in the commercial marketplace.

If we could create a national debate on these issues, and on a set of related issues that surround government funding of research and development, that would be a good start.

CHAPTER FOUR

Scientific Research in a Free Society: Some Reflections

Eleftheria Maratos-Flier

INTRODUCTION

Scientific research is a diverse enterprise involving investigation into areas ranging from cosmology to methods for making better test strips to check blood sugars in people with diabetes. Between the arcane and the mundane there is a continuum of innumerable subjects. Given the varied disciplines covered and the disparate implications of research in these areas for society, it isn't possible to provide one answer to the appropriateness and role of research in a free society.

Of course, a strict anarchist-libertarian would have a simple and ready answer which would be "no." Since taxation is force and theft is inappropriate, the state has no right to steal, even for a goal such as understanding and curing cancer, let alone determining the nature of the universe. Whatever the force of these arguments, we live in society governed by a state, and that state is not going to disappear tomorrow or even in the next century. This state funds research under a number of programs. Since libertarians would like to see the state evolve in the direction of more rather than less freedom, it is reasonable to ask whether or not the state has any role in a free society and, furthermore, to examine the potential effects of participation of the state in research on overall freedom. Does state involvement in research lead to more or less freedom, or is it without any particular effect? If the overall effect is neutral, is there any justification for government-funded research?

In considering these questions I define research as inquiry into either biological, i.e., life sciences, and physical sciences. I specifically exclude research in such areas as economics and sociology, as their nature is intrinsically different from that of biological and physical science and they are also beyond my expertise. When giving examples I largely emphasize the life sciences, as my vantage point is that of a biomedical researcher.

BACKGROUND

"Life sciences" is a very general term covering a broad range of inquiry from basic biology, to biomedical science, to clinically oriented medical studies. Research in the life sciences includes basic research, applied research, and development. These represent a continuum of inquiry that ranges from pure knowledge-seeking as an end in itself to the development of products (pharmaceuticals or medical treatments) intended to cure human illness or create improved agricultural products. The extremes are easy to differentiate, but at the borders the basic research may be difficult to distinguish from applied research.

Basic research in the life sciences might involve research in cellular development in flatworms or molecules synthesized by fungi. Applied research might involve regeneration of neurons or adaptation of fungal products to inhibiting cholesterol synthesis. At the development end, the research would involve identification of drugs that stimulate neuronal growth or drugs that reduce cholesterol in people, and the effect of those drugs on incidence of heart attacks and strokes. Areas of zoology, botany, or yeast biology might be viewed as purely basic, without implication for human disease, however there are unexpected spillovers. For example, a hormone important for regulating pigmentation of fish skin is an important regulator of eating in mice and man. When one then considers the areas of microbiology or immunology, or genetics or behavioral neuroscience, the distinction between biology and biomedical research becomes increasingly difficult. Even when the test subjects are bacteria or mice the implications for medical research may be significant. Similarly, research performed on human subjects can have significant implication for basic areas. Human Immunodeficiency Virus was discovered in people suffering from Acquired Immune Deficiency Syndrome; however, studies of the HIV as virus as opposed to a human pathogen are more in the realm of basic virology.

The range from basic research to applied research and development is best seen as a continuum with fuzzy borders. In the area of neuronal degeneration, basic research might involve genes that regulate development of cells flatworms; translating this research to understanding genes that regulate development of neurons in mice is still basic, but as the implications of this reasoning to Parkinson's disease or Alzheimer's disease are possible to visualize, it borders on applied research. Looking for factors that might stimulate neurons to grow in culture or in mice is even more applied. Looking for factors that can be used as drugs because they exert their effects without toxicity is applied research but also has aspects of drug development. Finally, the clinical studies performed to see if the drugs work as expected in humans are entirely part of the drug development process.

In the United States life science research occurs at a variety of sites, including academic centers, pharmaceutical and biotechnology companies, independent research institutes, and a variety of government sites. Most of the research performed at academic institutions is in universities with graduate programs or medical schools. Colleges without graduate faculty may have small research programs, but these programs represent a minority of overall life science research. Independent research institutes are usually privately funded institutions with or without university affiliations. From the standpoint of performing life science research they function much like departments in universities; senior scientists have advanced degrees and credentials similar to university faculty, the laboratories may train post-doctoral fellows and (if the institute has a university affiliation) graduate students as well. For both pharmaceutical and biotechnology companies life science research is essential to maintaining drug pipelines and pushing drug candidate molecules through the drug development process.

At the government level, life science research performance sites are almost always federal facilities and include the National Institutes of Health (NIH), the Center for Disease Control (CDC), and medical departments of some of the Veterans Administration Hospitals.

FUNDING

In 1998 all research and development expenditures in the United States were calculated at $227 billion, representing a record level of funding. Of these dollars 66 percent were provided by industry and 29.5 percent by various agencies of the federal government; 2.2 percent of the budget came from universities and colleges (this includes state funding to state universities) and 2.5 percent of the funds were derived from other sources, including state government and local governments and nonprofit institutions. This current distribution of funding has evolved over a five-decade period since the World War II, during which the federal contribution first increased sharply from 20 percent to 62 percent in the fifties and sixties and then declined to its current levels. In the decade preceding 1998, the contribution of the federal government fell in terms of constant dollars as well. Contributions from all other sources fell as the federal share fell, and have risen correspondingly.[1] Hence, while all categories (basic, applied research, and development) are at currently historically high levels of funding, the growth has occurred in the private sector, in the face of declining federal contributions, indeed the federal contribution to industrial research and development; in 1997 federal funds represented 14 percent of total financing (an unprecedented low).[2]

The decline in the proportion of federal funding to overall research was also seen for funding of academic research. The federal effort peaked at about 75 percent in the late sixties and then declined to 60 percent in the late 1980s. It has remained stable since. In 1998 a total of $26.3 million (about 10 percent of total research and development dollars) was spent for research and development at academic institutions. Of this 59 percent came from the federal government, 19 percent from academic institutions, 8 percent from state and local govern-

1. National Science Board, "Science and Technology in Times of Transition," chap. 1 in *Science and Engineering Indicators—2000* (Arlington, Va.: National Science Foundation, 2000), fig. 1-1 and table 1-3. *www.nsf.gov/sbe/srs.*

2. National Science Board, "U.S. and International Research and Development: Funds and Alliances," chap. 4 in *Science and Engineering Indicators—1998* (Arlington, Va.: National Science Foundation, 1998), fig. 4-5. *http://www.nsf.gov/sbe/srs/seind98/start.htm.*

ments, 7 percent from industry, and 7 percent from all other sources. This distribution varies between private and public academic institutions. Private institutions receive 72 percent of their research budget from the federal government, only 2 percent from state and local governments, and 10 percent from institutional funds. The balance is supplied from industry and from nonprofit research foundations. Eighty-three percent of the total federal funding derived from three agencies: the National Institutes of Health (NIH, 58 percent), the National Science Foundation (NSF, 15 percent), and the Department of Defense (DOD, 10 percent). Other federal support came from the National Aeronautics and Space Administration, the Department of Energy, and the Department of Agriculture. During the past decade the proportion of NIH, NSF, and NASA funding has grown, while funding from the DOD, DOE, and USDA declined. In the life sciences (excluding psychology) virtually all federal funding comes from the NIH (89 percent).

Life science research dominates academic research in terms of both funding and distribution of research space. In 1997 expenditures in medical and life sciences totaled $8 billion, compared to $3.4 billion in engineering, $2 billion in physical sciences, $1.4 billion in environmental sciences, and $1 billion in social sciences. Similarly, biological and medical sciences utilized 40 percent of research space, compared to 15 percent by engineering and 14 percent by physical sciences.

SUMMARY OF BACKGROUND

The contribution of the federal government is not static, but continues to play a role in research and development in the United States. Funding has declined over the past four decades and now represents less than 30 percent of total research and development funding in the United States. Although the proportion of research funding to academic institutions has declined from absolute peaks in the late sixties, the total has declined. Most of the decline is in federal contribution to industry. Federal support to academic research has also declined but federal support still dominates support to academic institutions.

GOVERNMENT AND RESEARCH FUNDING

To look at the specific question of the potential role of government in research in a free society, let me first consider the role of government. Approached from an anarcho-capitalist perspective the government has no role. Society should be structured so that individuals choose their protective agencies to provide necessary services. There is no government to perpetrate a drug war; there is also no government to provide defense against enemy nations. In an anarcho-capitalist society the answer regarding the role of government in research is simple. There is none; hence if libertarians are simply anarcho-capitalists, they have their answer.

However, between a society structured along the protective societies of anarcho-capitalists and the current representative, quasi-paternalistic, partially redistributive democracy that is the United States government there is a continuum. This includes the minimalist state that Nozick argues protective agencies would necessarily evolve into,[3] and the Jeffersonian government-that-governs-best governs-least state that the founders had in mind. Just as Catholicism is only one Christian dogma, anarcho-capitalism is only one libertarian theory. Clearly individuals may be considered libertarians and still favor a minimalist state or a Jeffersonian state. It is possible that some libertarians simply would like to see the state eschew

3. R. Nozick, *Anarchy, State, and Utopia* (New York: Basic Books, 1974), chap. 2.

prosecution of victimless crimes, permit any and all behavior among consenting adults, and cease redistribution of wealth among individuals.

What might such a state look like? In the minimalist format, the state would provide police and national defense services and support a judiciary. It would be in the business of protecting the life and property of its citizens against force and fraud. These services might be financed through a flat tax that had a constitutionally defined maximum to prevent the finances of the state from becoming too robust, thereby providing a system of checks on government activity based on financial limits. There are many things that the state would not do; in fact, it wouldn't do most of the things associated with the modern state. The state would not: be involved in the activity of proscribing any behaviors between consenting adults, regulate trade, banking, or stock trades, provide senior citizens payments for health care, accredit medical schools, proclaim Secretaries Day or National Dairy Week.

Is there justification for such a state to be involved in any way in the pursuit of basic research or applied research or development? The answer depends on whether or not the state can fulfill its obligations to its citizens by relying solely on markets that have evolved privately. There are at least two areas in which it seems unlikely that traditional markets, uninfluenced by government, would be adequate for providing necessary products to the state. One is in the area of national defense and the other is in the area of public health.

The necessary involvement of the state in at least assessing and possibly subsidizing research at all levels in the area of defense seems obvious. Customers in the market for national defense products are competing nation-states. Common sense dictates that they will develop and jealously guard from their enemies both offensive and defensive weapons. Ideally the minimalist state will be interested only in the best defensive weapons, and it is unlikely that such weapons will be available from suppliers in other competing nation-states in an open market. Some technology might be shared with allies; however, the acquisition of forefront technology requires internal research and development. It's easier to control what is internally developed. It seems likely, in order to maintain a competitive position in defending its citizens, that the minimalist state would be required both to assess and invest in the engineering and science research involved in the manufacture of defensive weapons.

The necessary interest of the minimalist state in the areas of public health research and environmental research is less obvious. These sciences have been, to a significant degree, co-opted by the political left and used as a rationale for proposing increasing state regulation, for anything from saving endangered snails to taxing high-fat foods to decrease rates of obesity in the adult U.S. population. While the politicization of these areas makes it difficult to objectively assess the significance of the claims and findings by scientists, nevertheless both public health issues and environmental issues can have direct effects on a populace.

It is completely consistent with libertarianism to claim that an individual's right to life entails a right to clean air that has an adequate oxygen supply. The extent to which human activities affect air quality should be of concern to a libertarian polity. If pollution leads to significant bodily harm, it follows that whatever agency provides protection for rights would have an interest in preventing the harm or offering a mechanism for obtaining restitution. The concern needs to be rational, and it can only be rational if the actual hazards can be defined. We can't say that the air needs to be clean, we need to ask how clean. In addition to being unpleasant, common pollutants cause small but measurable increases in asthma, bronchitis, and cancer. People who choose to live in cities like New York accept the small increase in risk. Over the past three decades, in United States cities air quality has improved, while disease risk has been reduced by laws regulating automobile emission controls. (If you doubt the effects of emission control on urban air quality, I recommend spending a few days in Athens, Greece, a city with inadequate emission controls.)

Meanwhile some environmentalists argue that emissions must be reduced further. Is this goal rational? Should people in Manhattan expect lower air quality than those living in Vermont? Answers require a risk-benefit ratio that can only be evaluated in the presence of data. The nature of the pollutants needs to be definitively known, the risk of the pollutants to health must be understood, and the cost of reducing the pollutants must also be known. The inquiry addressing these questions involves both environmental sciences and epidemiology. Environmental research would address the source of pollutants and their distribution in the environment, and epidemiology would address the prevalence and incidence of disease caused by pollutants. An excellent example of the difficulty in evaluating the problem of environmental pollutants can be found in a series of articles examining the "Arsenic Controversy" recently published by the Cato Institute.[4] Even the minimalist state would have an interest in funding such research, if such research were not being done privately, because the information might be essential for protecting rights.

When it is directed to addressing infectious diseases, the area of public health research might also be of legitimate interest to government. Does the government have any legitimate in-

4. "Special Report: The Arsenic Controversy," *Regulation* 24 (fall 2001): 42–54.

terest in infectious epidemics? I'll take it as a given that the state has no legitimate interest in noninfectious epidemics such as obesity. Although more than 50 percent of Americans are obese, and although this obesity has significant health consequences, it remains an individual problem related to the abundance of fat-inducing foods and a genetic predisposition to obesity. In contrast, infectious agents target individuals all the time during the annual cold and flu season and garner scant attention. This is because a cold or flu causes some inconvenience, but rarely complications. Companies lament losses in productivity and parents lament the days that their children stay home from school. Most people take it for granted that they'll get a virus or two in the winter but hardly expect the government to do anything about it.

However, the infectious world harbors nastier agents than myriad rhinoviruses causing colds. It harbors drug-resistant tuberculosis and malaria, and highly contagious, lethal viruses such as Llasa Fever and Ebola. Because infectious illnesses threaten life, to what degree should citizens expect that the state would be interested in protecting them from this kind of threat? Is the control of infectious epidemics a legitimate goal of the state? Infectious epidemics aren't purely a private concern. To the degree that they are infectious and potentially lethal, society has an interest in trying to identify and control epidemics. If this is a legitimate goal, then the state needs to ensure that there is adequate knowledge about the infections that might cause such epidemics.

It is possible that information for both environmental and public health questions would come from privately funded sources. To the extent that they are concrete problems such as smog, and have health consequences that are not to difficult to identify, as in asthma, citizens may well decide to fund agencies privately out of concern for their own health. To the extent that the problems are obscure, not immediate, either in time or geography, it may be necessary to centrally fund research that will address the problems.

Furthermore, infectious agents, some of which pose serious intrinsic threats, can also be used as weapons. The recent public health chaos over the possible threat of anthrax as a bioterrorism agent serves to highlight the need for current information in formulating policy decisions. Thus far the Center for Disease Control has promoted conflicting policies with regard to prophylaxis of individuals potentially exposed to anthrax. However, in what might be considered the "molecular era" of medical research, little work has been done on anthrax over the past twenty years. Other than a poorly coordinated, federally funded effort at generating an anthrax vacine for military personnel, there have no notable attempts at developing new treatments. In effect, anthrax, which poses little natural threat, is an "orphan" disease. The fifteenth edition of *Harrison's Principles of Internal Medicine*[5] has two pages devoted to anthrax and sixty to Human Immunodeficiency Virus. Physicians, the CDC, local public health departments have little information to go on. Local public health departments are also ill-prepared. Neither the public nor private sector provided adequate preparation, and had there been a widespread release of anthrax, hospitals and medical personal would have been significantly challenged to provide adequate care to those infected. It is possible that, absent the events of September 11, neither the private firms nor the state had any motivation to research anthrax, as any long-range view that it might be used for bioterrorism was considered

5. *Harrison's Principles of Internal Medicine*, 15th ed. (New York: McGraw-Hill, 2001), 914–15 and 1852–1912.

remote. However the reasons for considering anthrax as a weapon were credible and anthrax and other bio-weapons represent an area where defense interests and public health interests become confluent.

For example, if antibiotic-resistant anthrax were used as a weapon, it might be possible to treat anthrax with antibodies that interfere with the action of anthrax toxins. No such antibodies are currently available. In such a scenario, anthrax vaccine might also be effective. Hence it seems appropriate for the state to fund research in the area to ensure that efficacious prophylaxis and treatment are readily available. This is a problematic conclusion as it is not clear that future efforts would be more efficient than initial efforts at obtaining vaccine for troops. Also, it is important to note that this conclusion is not as firm as the conclusion regarding the role of the state in supporting research in traditional defense. Health care-related research may be inhibited or prohibited by regulations of the Food and Drug Administration or other government agencies. To the extent that this is the case, it is important to address these impediments and remove them before concluding that the support must derive from the state.

SUMMARY

The minimalist state will need to fund research to the extent that citizens expect it to protect them. Given that the market for national defense items is likely to be limited to other competing nation-states, the federal government will have to fund research into appropriate products for its own use. In the case of environmental and contagious public health issues, some or all research may be privately funded and the appropriate role of the state may be to ensure that sufficient research is done.

BEYOND BASIC OBLIGATIONS

If it is acceptable for the minimalist state to fund research intrinsic to protective functions that the citizenry views as contractual obligations, provided that such research isn't going to be financed privately, is it ever acceptable or desirable for the state to fund research outside the market, beyond what it needs to do to adequately fulfill its mission? I don't think that this is a question with a simple "yes" or "no" answer. For a pure minimalist state, the answer would seem "no." If the sole function of the state is to protect citizens against external threats, why would the state need to fund research outside the areas described above?

However, the minimalist state is a theoretical construct. The United States isn't a minimalist state. Assuming a relatively stable world, I doubt that many people would predict that the United States will evolve rapidly into a libertarian minimalist state over a few decades. It's not practically possible to predict how the world or our nation will evolve over the next century. Most of the predictions made in 1900 were wrong, so it is worth asking the question about the appropriate role of the state in the context of present reality. Our state, a representative democracy that engages in some redistribution of wealth, exists in a world with other states, most of which are less democratic and engage in greater degrees of redistribution of wealth. In this context should the state support research? Should the support be limited to certain kinds of research? Is the support of research likely to increase or decrease personal freedom? It is beyond the scope of this essay to explore the nature of personal freedom, so I will be conventional and say that personal freedom is the ability of an individual to be a project-pursuer,[6] to act in a way that fulfills individual desires and goals as long as, in the process, one does not engage in force or fraud. Wealth typically correlates well with personal freedom in that the wealthy find it easier to pursue projects than the poor. Similarly, it is easier to be a project-pursuer in a wealthy society than in a poor one because wealthier societies are freer and less susceptible to tyranny. To the extent that discovery and invention have improved health and productivity, scientific research increases wealth. As economics is beyond my area of expertise, I refer the reader either to a serious libertarian approach to the subject or alternatively to P. J. O'Rourke.[7]

The United States spends more money, both private and public, in the pursuit of research and development than any other nation, and certainly produces the highest percentage of publications in every scientific field (although on a per capita basis Switzerland, Sweden, Israel, and Denmark are more productive).[8] In terms of a journal-impact factor, of the top five journals four are published in the United States (*Cell*, *New England Journal of Medicine*, *Science*, *Journal of Experimental Medicine*) and one is British (*Nature*). The combined dominance of the United States and Britain in science has led to an inexorable shift to English as the main language of science. As of 1997 95 percent of papers actually cited by another paper were published in English, compared to 83 percent in 1977.

The U.S. economy is the world's largest and citizens enjoy

6. L. Lomasky, *Persons, Rights and the Moral Community* (Oxford: Oxford University Press, 1986).

7. P. J. O'Rourke, *Eat the Rich* (New York: Atlantic Monthly Press, 1998).

8. T. J. Phelan, "Evaluation of Scientific Productivity," *Scientist* 14 (2000): 39.

one of the highest standards of living. To what degree is this success related to the success of technological industries? Causality seems hard to prove, but it is clear that both prosperity and productivity are associated with the high-technology products. Analysis of the global marketplace in technology indicates that the United State is the leading producer of high-technology products and is responsible for one third of these products. High-technology industries include computers, pharmaceuticals, communications equipment, and aerospace. During the 1990s, the United States has gained market share in all but the aerospace industry. Overall high-technology industries are important in the generation of wealth.[9]

Assuming that a specific activity does not result in decreased freedom and is not meant to substitute for activities already privately supported, it does not seem to me to be antithetical to libertarian principles to have the government support certain types of research if the intent of the activities is to increase knowledge, wealth, and, ultimately, freedom.

How might state funding of research *decrease* freedom? To consider this question we first need to examine the nature of research. Research is an activity performed by scientists, usually in a laboratory setting. The scientists are there voluntarily. If the research involves human subjects, an extraordinary number of precautions are taken to ensure that the subjects are also there voluntarily and have given informed consent. When the subjects are vertebrate animals, care is taken to ensure that the protocols used are appropriate and that the researchers minimize pain. If research is performed in an academic setting, the results of the research belong to the laboratory performing the

9. National Science Board, "Industry, Technology and the Global Marketplace," chap. 7 in *Science and Engineering Indicators—2000* (Arlington, Va.: National Science Foundation, 2000).

research. It is the responsibility of the lab chief to ensure that the data are reliable and to disseminate the results. In an academic setting, any potential patents name the investigator as inventor. Even if the government has funded the research through an agency such as the National Institutes of Health, it eschews ownership and allows the institution to own the patent. A significant portion of state-funded research (most of the portion not used to fund defense) is a voluntary activity engaged in by scientists at academic institutions. Although government-funded, the ethos of the research taking place at academic institutions comes from the academic environment. Funds are derived from the state, but participation as a scientist or as a subject is voluntary.

The possible concern for coercion is that state funding of research generally means that the funds come from taxpayers. A libertarian who feels that any payment of taxes is a coercive activity cannot possibly condone government funding of basic research. A libertarian who feels that some taxation is permissible (be it in the form of a flat tax), because of the benefits of living in a free and wealthy society, might be convinced to support basic research on the basis of future returns. The total contribution of the federal government to all research and development is $66.9 billion. This represents 4.0 percent of the $1.652 trillion federal outlays of 1998. The amount spent at academic institutions was 1.5 percent of the total budget.[10] Compare this to the total 35 percent of the budget spent on the largely unjustified transfer of wealth from younger individuals to older individuals in the form of Social Security and Medicare payments. A society that eschews enforced transfers

10. U.S. Census Bureau, *Statistical Abstract of the United States: 2000*, table 532.

of wealth can well afford to fund research on the income from a modest flat tax.

THE PROBLEMS WITH A
MARKET FOR RESEARCH

Ideally, funding for research should come from private sources. For any particular area of inquiry, the nature of the research ranges through a continuum of basic to applied. The more basic the question, the lower the likelihood of private funding. Why? It is because the product of basic research is information, which only has market value if it can be sold or traded for something else. For example, the information "this is the gene that codes for erythropoietin (EPO), a hormone that stimulates red blood cell formation" has value because EPO can be sold (and indeed is now sold) to people who are anemic because of low EPO levels. The information "this is the gene that encodes for the hormone, melanin concentrating hormone (MCH), a hormone that causes fish color to lighten" has no economic value because there's no known market for treating fish with expensive peptides to change their skin color. Basic research is pursued with the sole intent of understanding a biological or chemical or physical process and without any specific intent to produce products. Although ultimately the research may have broad applied implications, these may have never been envisaged by the scientists making the original observations.

Let's consider the story of a well-known molecule, insulin. As potential users of medical care, most us living in Western democracies take for granted that if we develop diabetes and require insulin we will be able to go to the drug store and readily purchase insulin. Most of the potential users are un-likely to consider the source or the history of the product, which illustrates the interaction of basic and applied research and, I think, illustrates the problems of expecting market forces to fund basic research.

Insulin was discovered in early 1920s as part of an applied research effort. It was already known that removal of the pancreas in dogs led to a syndrome of high glucoses that looked similar to human juvenile diabetes. The effort, which was not the first to isolate insulin, was initiated by a surgeon, Frederic Best, who later (if somewhat controversially) won the Nobel Prize.[11] Insulin was first introduced as a life-saving drug isolated from either beef or pork pancreases in 1922. Insulin worked well enough, at least in juvenile diabetics. Eli Lilly and Company was the main manufacturer in the United States. Beef pancreases were more abundant but pig insulin, which differs from human insulin by only one amino acid, was considered a better product because it was less likely to produce allergic reactions. Between 1922 and the 1970s the major research efforts of Lilly and the European manufacturers of insulin (Novo and Nordisk) were aimed at refining the purification methods and the formulations to create short, intermediate-acting, and long-acting products. Insulin may have provided a large profit for these companies, yet none were involved in further studies aimed at understanding the molecule.

Still, insulin was the subject of basic research. In 1953 Fred Sanger reported the sequence of human insulin.[12] This was a historically important finding because what is now obvious, that specific proteins have specific structure of amino acids

11. M. Bliss, *The Discovery of Insulin* (Chicago: University of Chicago Press, 1984).
12. F. Sanger, "The Arrangement of Amino Acids in Proteins," *Advanced Protein Chemistry* 7 (1952): 1–67.

from end to end, was then a subject of debate.[13] In the 1960s, three groups, one in the United States, one in Europe, and one in China, reported on the chemical synthesis of insulin, demonstrating that this fairly long protein could be synthesized through a series of chemical reactions in a test tube. The chemical synthesis of insulin was a scientific climbing of Mount Everest and had about equal commercial use because the synthesis was too expensive to be considered by industry. For a brief period, in the mid-to-late 70s, physicians involved in the treatment of diabetes voiced concern that by the year 2000 there would be a global shortage of insulin. As the number of insulin-treated diabetics grew, it seemed unlikely that the material from slaughterhouses would be adequate to supply the demand.

Between the discovery of insulin and the synthesis of the peptide, other discoveries were made. The structure of DNA and the nature of the genetic code were discovered in 1952, followed by a couple of decades studying DNA and genes, mostly in one bacterium, E. coli, a resident of human gut. A lot of this work focused on subjects that seem arcane, even to me. How much do you want to know about mutations in DNA factors regulating production by bacteria of an enzyme that breaks down the sugar galactose? The research had no commercial applications and the scientists were unlikely to win public acclaim by discovering the cure for cancer. Even if you've studied biology in college you may not remember that Jacob and Monod won the Nobel Prize in 1965 for their discovery of the lac operon, a regulator of a bacterial gene.

Meanwhile, much information about bacteria was being discovered. Because of their nature, bacteria are susceptible to

13. J. Darnell, H. Lodish, and D. Baltimore, *Molecular Cell Biology* (New York: Scientific American Books, 1986).

invading bits of DNA, and to deal with foreign DNA make enzymes that degrade it. However, the bacteria have to distinguish between their own DNA and invading DNA. Bacteria accomplish this through molecules known as "restriction enzymes." The earliest reports on the enzymes, dating back to the late 1960s, were published without much fanfare. While the information might seem as arcane as the information on lac operon, these restriction enzymes form the basis of the biotechnology revolution.

Scientists, notably Arthur Kornberg, spent years studying the mechanisms by which DNA is synthesized and degraded, in the process discovering a number of enzymes that could be used to manipulate DNA. For example, scientists can ligate pieces of DNA cut with a restriction enzyme to other pieces of DNA or make copies of DNA that they already have. Using the variety of enzymes derived from examining bacteria, scientists can now identify a gene, make a probe to study it, modify it, devise a way of making the product of that gene, and ultimately engineer a bacterium or a yeast cell that incorporates the gene of interest in its genetic material and makes the product protein. Alternatively they can engineer a mouse that makes the peptide they are interested in, in the chosen tissues. Nature puts insulin production in the pancreas, but a scientist could place it in the stomach or in the testicle. If you're a scientist and you're feeling particularly silly, you can make a rabbit that glows green.

One of the first genes to be manipulated in this way was the insulin gene. In the mid-1970s a group of research scientists led by Walter Gilbert at Harvard Medical School described the identification of the gene for insulin and the production of insulin by a bacterial clone.[14] Within a few years scientists had

14. L. Villa-Komaroff, A. Efstratiadis, S. Broome, P. Lomedico, R.

generated recombinant bacteria capable of manufacturing insulin, and within a few more years part of the world's insulin supply was coming from large fermentors housing billions of bacteria busily making a human protein entirely foreign to the bacterial chromosome. Today the vast bulk of the world insulin supply is manufactured through the use of genetic engineering and what is sold at the drug store is "recombinant human insulin."

It is not surprising that once Gilbert isolated the gene for insulin, pharmaceutical companies would imagine how to exploit the discovery. However, the 1978 paper reporting on the synthesis of insulin by E. coli was based on almost three decades of research without commercial applicability. Kornberg has commented, "No industrial organization had, or ever would have, the resources or disposition to invest in such long-range, apparently impractical programs."[15]

Basic research, in its pure form—the investigation of processes aimed solely at understanding the process—is an activity that cannot be sustained by a market. The study of DNA replication in bacteria yielded interesting information on the nature of bacteria, but the interest was limited to a small audience, other scientists interested in DNA.

What's the market for the results of basic research that have no possible commercial information in the foreseeable future? It has enough value to be published in journals, but science journals do not pay royalties to the authors. Publication confers prestige, especially if other scientists like the work and cite

---

Tizard, S. P. Naber, W. L. Chick, and W. Gilber. "A Bacterial Clone Synthesizing Insulin," *Proceedings of the National Academy of Sciences USA* 75 (1978): 3727–21.

15. A. Kornberg, "Support for Basic Biomedical Research: How Scientific Breakthroughs Occur," *The Future of Biomedical Research*, ed. C. E. Barfield and B. L. R. Smith (Washington, D.C.: American Enterprise Institute and Brookings Institution, 1997).

it. But the prestige has a limited audience; it is other scientists that understand the work. Information that can't generate commercial interest is information that has no market value if you can't make a drug, build a faster computer or a more energy-efficient furnace.

Intrinsic to the nature of basic science is that it is pursued without any clear-cut awareness of the market value of the information. Scientists doing basic research today have motivations not dissimilar to scientists in the time of Aristotle. Aristotle considered nearly every aspect of the world that he observed[16] from the vantage point of a scientist, because it was there to be catalogued, not because of the marketing potential of what he had catalogued. Value, to a basic researcher, is a publication in *Nature*, *Cell*, or *Science*. A patent and royalties might be nice, as they might pay for the college education of the scientist's offspring, but they aren't the goal. Although many scientists dream occasionally about doing something worth a Nobel Prize, most work with enthusiasm knowing they are never going to win one. Not only are the products of basic research not subject to value in a conventional market, but those laboring at basic research tend to ignore the rewards of the conventional market. Many scientists working in an academic institution could easily increase salary by 50–100 percent working for a biotech or a pharmaceutical drug company, yet excellent scientists stay in academics, with lower salaries and the problem of worrying about grant funding in exchange of the freedom of pursuing basic questions.

Even if basic science produces products that are of questionable immediate market value, why not rely on concerned citizens funding basic research privately? Concerned citizens band together to form organizations that fund disease-related

16. Aristotle, *Selected Works*, trans. H. G. Apostle and L. P. Gerson, 3d ed. (Grinnell, Iowa: Peripatetic Press, 1991), 1–27.

research such as the American Diabetes Association, the American Heart Association, and the March of Dimes. Generally the total effort funded by these organizations is small, and although some basic studies are funded, much of the research is aimed at finding a cure. Private philanthropy aimed at funding basic research is unusual. The Howard Hughes Medical Institute is unique, both in its method of funding individuals rather than projects and in contributing to biomedical research. In 1993 it contributed $268 million to biomedical research. Although this contribution is significant, coming from a single organization, it represents only about 1 percent of total research funding. Would private agencies contribute more if the government contributed less? This seems unlikely, given the fact that this source of funds has been stable at 7 percent of academic research and development funding for over three decades.

Investment by biotechnology industry and pharmaceutical companies in research is significant, but it tends to be directed at research where at least a product can be envisioned. I referred to the gene for the fish hormone, MCH, as an example of a bit of information that has no commercial value. Indeed although the hormone was identified a British group in 1983, pharmaceutical companies had no interest in this substance until 1996 when my research group demonstrated that in mammals, MCH stimulated feeding behavior. Going from a fish hormone mediating color change to a mammalian neuropeptide important in appetite and obesity changed the interest of industry, and I am now aware of a number of companies with MCH projects. However, in 1983 no company could have convinced investors that MCH was a project worth working on.

Investment by private individuals in basic research without a disease focus through charitable organizations is unlikely to happen. As discussed above, the benefits of basic research may only be seen long-term. Most individuals don't have enough understanding of scientific process to grasp the long-term consequences of basic research. It is depressing to note that only 11 percent of Americans can define the term "molecule" and only one in five can provide a minimally acceptable definition of DNA. Only 50 percent reject the statement that "the earliest humans lived at the same time as the dinosaurs." Astoundingly only 48 percent know that the earth goes around the sun once a year![17]

Finally, with regard to government funding of basic research, I believe that funds should be directed to universities and institutes outside the federal government. The National Institutes of Health supports some intramural research of generally high quality. However, there is nothing special this research. As the subject areas explored at the NIH are explored elsewhere, there seems to me no justification for directing government funding at government institutions. A notable exception may be the Center for Disease Control, particularly in its role in identifying and monitoring infectious epidemics. This public health role may be hard to duplicate in a private setting.

SUMMARY

The products of basic research do not necessarily have market value and the individuals that pursue basic research do not necessarily act to maximize their own economic circumstances. The results of basic research may have long-term market value, however. Private investment is usually directed at

17. National Science Board, "Science and Technology: Public Attitudes and Public Understanding," chap. 7 in *Science and Engineering Indicators—1998* (Arlington, Va.: National Science Foundation, 1998). *http://www.nsf.gov/sbe/srs/seind98/c7/c7s2.htm.*

knowledge that has short-term value. Neither industry nor private charity has a record of funding research for the sake of research. Funding as a potential government activity is reasonable if it is aimed at activities for which the government has obligation to citizens (such as defense). Funding is also permissible if it is directed at research that would not be supported by market forces and if it is aimed at increasing freedom and wealth.

THE DOWNSIDE

Assuming that one agrees that in a free society it is reasonable for the government to fund basic research in the life sciences and the physical sciences, with the constraint that this research would not be otherwise funded, the potential negative consequences of actually enabling funding need to be considered.

One negative is that a certain amount of money would be wasted. This seems to be a given in any human activity. Even pharmaceutical companies, where there is constant pressure to create products and watch a bottom line, fund projects that never yield results. The extent to which money is wasted will depend both on the total funds available and the level of funding, and on the mechanisms by which the funding is distributed. An analysis of the NIH review process, which funds "projects," is beyond the scope of this chapter. However, it is important to note that at present 25–30 percent of investigator-initiated grant proposals are funded after a peer-review process. The process is not perfect, however the process is competitive and there is a serious attempt to fund the best science.

Another negative is that science will be politicized. This certainly happens. Jerome Groopman has written an excellent article on history of the war on cancer.[18] When Richard Nixon

18. J. Groopman, "Annals of Medicine: The Thirty Years' War: Is There a Better Way to Fight Cancer?" *New Yorker*, June 4, 2001.

declared a war on cancer in 1971, he was responding to lobbying efforts by the philanthropist Mary Lasker and the pediatric oncologist Sidney Farber. As Groopman points out, the war has not been won, despite significant funding. The initiatives were criticized at the time by a number of scientists who felt that a cure would not come by directive. Dr. Francis Moore, a medical historian, referred to the "law of unintended consequences in scientific discovery." Much of discovery happens by chance, which is the point of funding basic research in the first place. Dr. Moore pointed out that anyone interested in scientific discovery would not have supported the work of John Enders when he attempted to grow mumps virus. No one would have predicted that this work would lead to the process that produced polio vaccine. Yet we see examples of politicized funding frequently. The efforts directed at increasing breast cancer and HIV funding for finding cures are political efforts by interest groups. Scientifically there is no reason to expect that major advances in these diseases will necessarily come from directed research efforts as opposed to a serendipitous discovery made by a scientist examining an unrelated process.

Other types of political pressure also emerge. Special efforts are aimed at increasing the rates of participation of underrepresented minorities in research, however private initiatives are hardly immune to this pressure. In 1994 the Hughes Institute made serious attempts to increase the participation of women and minority groups in its research programs.[19]

There is also the potential of funding fads. A recent NIH initiative involves funding of research directed at alternative therapies, which might be useful to the extent that rigorous clinical studies are aimed at examining real effects of alternative therapies. The purveyors of alternative therapies seem to

19. K. Y. Kreeger, "Hughes Institute Moves to Bolster Female and Minority Participation," *Scientist* 8, no. 7 (April 4, 1994): 1.

be either unable or uninterested in performing such studies. If the money is used by alternative therapists to show borderline psychological benefit of such therapies, it is likely that the money will have been wasted.

Finally, a significant risk of permitting government involvement in research is that there are and will continue to be attempts by institutions and individuals to acquire support outside the peer-reviewed funding process. It is well known that line items can be added to appropriations budgets to fund congressmen's pet projects. Such funding, which is hard to identify, has been used to fund research projects at medical centers and universities. It is difficult to see this as anything other than pork-barreling, which I believe is antithetical to the aims of a free society.

These potential problems indicate that the process of funding needs frequent review and that its necessity cannot be assumed. The proportional contribution of federal funding to research has declined since the peaks of the 1950s and 1960s, and we can't assume that some level will always be needed. Depending on the nature of discovery, the need for funding may decline as more processes are understood. Alternatively, areas of current questions in basic research may open other areas, so the need for basic research may be necessary for a long time to come.

INDEX